Rational Politics

Decisions, Games, and Strategy

Rational Politics

Decisions, Games, and Strategy

Steven J. Brams

New York University
Department of Politics
New York, New York

ACADEMIC PRESS, INC.

Harcourt Brace Jovanovich, Publishers

Boston San Diego New York
Berkeley London Sydney
Tokyo Toronto

Copyright © 1985 by Academic Press, Inc.

All rights reserved. No part of this publication may be reproduced or transmitted in any form or by any means, electronic or mechanical, including photocopy, recording, or any information storage and retrieval system, without permission in writing from the publisher.

Academic Press, Inc.
1250 Sixth Avenue, San Diego, CA 92101

United Kingdom Edition published by
Academic Press Limited
24–28 Oval Road, London NW1 7DX

ACKNOWLEDGMENTS

The author gratefully acknowledges the following publishers for use in *Rational Politics* of portions of the works cited below.

The Free Press, a Division of Macmillan, Inc., from *Game Theory and Politics*, copyright © 1975; and from *Paradoxes in Politics: An Introduction to the Nonobvious in Political Science*, copyright © 1976 (in Chapters 3, 4, and 7)

Yale University Press, from *The Presidential Election Game*, copyright © 1978 (in Chapter *8)

Consortium for Mathematics and Its Applications (COMAP), from *Spatial Models of Election Competition*, copyright © 1979 (in Chapters 2 and 8)

The MIT Press, from *Biblical Games: A Strategic Analysis of Stories in the Old Testament*, copyright © 1980 (in Chapter 2)

LCCCN 89-045887
ISBN 0-12-125455-0

Printed in the United States of America

89 90 91 92 9 8 7 6 5 4 3 2 1

Preface

"Rational politics" may sound like a contradiction in terms, at least in some situations. The thesis of this book, however, is that the choices of political actors, if not always coolly calculated, are generally the best they can make, given their goals and the situations they face.

To illustrate the rational calculations of political actors, I start with the Bible and then analyze several modern real-life cases, ranging from the 1962 Cuban missile crisis to the current superpower arms race, from Watergate to the recent Polish crisis, from citizens' voting in elections to legislators' voting in Congress to countries' voting in the European Community Council of Ministers. I explicate these examples by building rational-choice models, which serve as the foundation of the analysis.

By *models* I mean simplified representations of political phenomena and processes that abstract the essential features of what one wishes to study. They are grounded in decision theory, game theory, and social-choice theory, which I introduce gradually throughout the book. My purpose in using elementary tools from these theories is to add clarity and rigor to the analysis, which is deductive in nature but inspired by observations of the political world, especially the strategic calculations of its players. I avoid certain technical details to highlight the important ideas.

Still, one cannot breeze effortlessly through this book. On the contrary, deductive arguments require considerable thought and reflection to absorb and comprehend. I believe the reward in greater understanding justifies this effort, which is eased by extended discussion of the more subtle points in the arguments, the use of descriptive aids in the many figures, and a glossary at the end of the book.

I have borrowed from some of my previous writings, both published and unpublished. The organization of the material is new, however, and its range is much broader than anything I have previously attempted. Also, I have updated all material to reflect both new research and recent events.

The major innovation of this book lies in its unified presentation of politics as a rational human activity, from ancient times to the present. Politics may appear bizarre, even topsy-turvy, on occasion, but some probing usually reveals that actors are striving to achieve certain ends. Models of rational choice help one determine exactly what courses of action satisfy these ends. Such courses of action may then be compared with actual choices and the empirical validity of the models thereby assessed.

I also analyze political institutions, such as voting and apportionment systems; as nonhuman actors, these systems do not, of course, make conscious or purposeful choices. Nevertheless, because they are human inventions designed to achieve particular ends, their theoretical properties and empirical effects can be considered in light of what they were intended to do.

Emphatically, rational politics does not imply good and desirable ends—only well-chosen means to achieve them. Normative judgments about both the means and the ends of rational politics, however, can and should be made. I indicate what consequences derived from the models I deem commendable or lamentable throughout the book and encourage readers to do the same. The models should help them distinguish, in terms of their own values, better from worse aspects of rational politics and thereby more carefully formulate and better defend their own views.

Students of political science—whether their interests lie in American government or international relations, in political theory or comparative politics—will find multifarious connections to the central theme of this book: the fundamental rationality of political thinking and political action, particularly in complex strategic situations that are hard to delineate, much less understand, without the aid of models. *Rational Politics* will prove most useful, I believe, for those instructors who want to stress the overall coherence of politics, perhaps in an introductory political science course but more likely in courses on political economy, public choice, modern political analysis, political methodology, empirical theory, or formal modeling. The book can also be used as a supplementary text in survey courses on different approaches to, or different methods of inquiry in, political science. Finally, *Rational Politics* is intended for those who simply are passionate about comprehending the world, in which politics—for better or for worse—is inescapable.

Several publishers of previous books have graciously consented to allow the use of material in *Rational Politics,* as acknowledged on the copyright page of this book. I am grateful for these permissions to use previously published material, which I have frequently condensed or combined with other work in the designated chapters to suit the needs of this book. For example, I have excluded some of the more technical material as well as the exercises and answers in the COMPAP teaching monograph, but I have included some new material on the 1980 and 1984 U.S. presidential elections. Generally speaking, the adaptations and juxtapositions are meant to mirror more closely the theme of politics as a rational human activity (see Chapter 1).

I am deeply grateful for the detailed comments of two perspicacious reviewers of the book manuscript, Russell Hardin of the University of Chicago and Daniel Sabia of the University of South Carolina, although I have followed only some of their advice. They should be absolved of responsibility for errors and misinterpretations that remain, as should my coauthors of a number of articles and papers from which I have also adapted material. Joanne D. Daniels, director of CQ Press, and Nola Healy Lynch, developmental editor, were supportive from the start. The merits of this book as a textbook owe a great deal to their appreciation of the needs of students and their editorial suggestions to meet those needs without compromising the book's intellectual content and integrity. Nancy Fernandez did her usual excellent job of typing the manuscript; the financial assistance of the C. V. Starr Center for Applied Economics at New York University for typing is appreciated.

Finally, heartfelt thanks are due to my family, Eva, Julie, and Michael, for their sympathetic understanding, especially when I became a bit tied up by the writing. It is no denigration of their own rationality to say that it was accompanied by cheerfulness and good humor.

Contents

Figures and Tables

Figures

Tables

1 The Study of Rational Politics

Politics is a subject difficult to pin down, though discussions of politics have a long and venerable history. No less a figure than Aristotle propounded his views in a book called *Politics* more than two thousand years ago. Of course, politics certainly predated Aristotle, manifesting itself, for example, in events reported in the Bible. To further complicate the subject, the politics of the twentieth century is vastly different in scope and substance, if not in form, from the politics of the ancients. Thus, it is not unreasonable to be perplexed about the meaning of politics.

There is, however, a core to politics that usefully distinguishes it from other human activities. Political relationships are characterized by *conflict,* which may be about any variety of matters, from civil rights to education, from agriculture to the environment, from the international economy to nuclear deterrence. What makes conflicts political is that they are not purely private matters but have a public dimension: they impinge on and involve other people besides the protagonists. The origin of the word *politics* reflects its public aspect—the Greek word *polis* means city or state, which is also the root of *politēs* (citizen) and *politeia* (government or constitution). In other words, politics by its very nature embraces a larger community—the public, or the citizens of a state—and how it is governed.

If politics is about *public* conflicts, then conflicts in the family, the office, and other more or less private arenas are outside the domain of politics, at least as the term will be used here. But this does not mean that politics is solely concerned with matters of government or state in which public participation is officially sanctioned. To limit politics to controversies connected with governing by official bodies or representatives is to take too restrictive a view. It shuts out conflicts that are indisputably political in the sense of affecting the lives and fortunes of many individuals, even the well-being of nation-states. Some multinational corporations, for example, are more powerful than some countries, if one's criterion is the impact they have on political leaders as well as ordinary citizens.

To exclude the quarrels of home, office, and other small-scale arenas but to include certain battles set in motion outside government requires a definition that avoids indelible boundaries yet establishes a general area of inquiry centered around conflict that affects the public. Different kinds of conflict will be distinguished later with the aid of game theory, including conflicts that never break into the open because they are deterred by threats.

The study of politics, even limited to the discussion of certain kinds of conflict, still requires a focus if it is to be understood as a coherent whole rather than a set of myriad details. The focus here will be on the analysis of *rational politics,* which can be defined as politics involving the *calculation of advantage and disadvantage by rational actors in situations of conflict, resulting in choices whose consequences affect significant numbers of people and the actions of governments.*

To act rationally means, in general terms, to choose better alternatives over worse ones. Since the ranking of alternatives depends on an actor's goals, rationality cannot be divorced from the goals that give it direction.

Political actors have all manner of goals. Some hunger for fame, whereas others prefer quiet, behind-the-scenes manipulation; still others seek mastery of a public policy issue or try to exploit such an issue because they have an ideological ax to grind. To the extent that actors achieve their goals as efficiently and effectively as possible, they are being rational. This is true even if they are not fully conscious of their goals but act *as if* they have them—that is, in a manner consistent with attaining certain goals.

Thus, to be rational is to strive for what one desires—or at least to act as if one were pursuing some end. But it is not simple to determine exactly what a rational course of action is that satisfies these desires or ends in a particular situation. Also, it is no mean task to check that a certain course of action was actually selected in order to test empirical hypotheses that are derived from a rational-choice model.

As will be demonstrated, the rationality assumption explains political behavior well, including that which seems on occasion paradoxical. Indeed, insofar as rational-choice models illuminate nonobvious aspects of politics, the use of these models seems justified.

To be sure, the models developed throughout this book differ substantially in the assumptions they make. Nevertheless, all share the same structure: consequences are deduced from assumptions, which makes the models deductive. To the degree that these consequences can be interpreted in terms of real-world politics, the models offer a palpable understanding of such politics, particularly of why conflicts break out and how they may be resolved or controlled.

In contrast to deduction, induction involves searching for patterns or regularities directly in the empirical data at hand rather than using data to test consequences derived from a model. Induction does not presume starting from an explicit logical structure that is rooted in relatively few assumptions. Instead, the inductive study of politics begins with the empirical description of the actual choices of political actors, the behavior of political institutions, and the like. Such description and the generalizations drawn from it, however, explain little if it is not clear why some alternatives were selected over others. For this reason it is usually insufficient to explain political behavior by observing it: it does not often speak for itself, no matter how carefully it is scrutinized or measured. Some prior assumptions are generally helpful and sometimes necessary for sorting out and making sense of the confusions and contradictions of political life, especially in trying to account for the choices political actors make that appear to violate common sense.

A deductive structure facilitates logically relating political choices to each other through a set of hypotheses, which are simply the consequences of a model interpreted or operationalized so as to be testable. The hypothetico-deductive method thus provides an economical means for going from an abstract deductive structure to a testable theory; it has become the hallmark of the natural sciences, which until the eighteenth century were largely descriptive and typological.

The hypothetico-deductive method is still not common in the social sciences. But there is no reason why, in principle, the study of politics should be impregnable to this method. In fact, it is our rationality that makes our activities intelligible. The assumption of rationality is now generally accepted in the field of economics, at least for explaining the behavior of individuals and households in microeconomic theory.

Political science has still not generally accepted this assumption, but not because political actors are less rational than economic ones. Rather, there is a paucity of rational-choice models—wedding the rationality assumption to the hypothetico-deductive method—to explain their behavior. This situation is changing, however, as it becomes more and more evident that political actors also think carefully about the options they have available and choose those that best enable them to achieve their goals. In short, the rationality assumption is by and large a realistic one.

It would be strange indeed if this were not the case, particularly when the decisions of actors have momentous consequences, for themselves as well as for the public. The only behavior that might be construed as irrational is random behavior, but even randomness can be shown to be optimal in certain situations and so may not be devoid of purpose. On the contrary, being utterly predictable may be utterly foolish in politics.

Politics, of course, connotes more than theoretical calculation, whether of the sinister variety associated with machination and subterfuge or the ennobling variety associated with courage and heroism. It is also about the actual decisions of citizens and elites, whose rationality may be severely circumscribed by many factors. Lack of information, for example, may significantly impede the ability of actors to make strategic calculations, as in the "fog of war."

Not only may information be imperfect, but political actors also may be constrained by limited resources, poor communication, their inability to make complex calculations, and a host of other real-life mitigating factors. These constraints, however, do not negate the actors' rationality. In fact, when one makes strategic choices in a murky or recalcitrant political environment, one is likely to be *more* prudent than if the obstacles are known and therefore can be anticipated.

The postulation of rational behavior is a powerful engine for systematically generating hypotheses from a small set of explicit assumptions. In addition, since optimal strategies are sometimes far from obvious, the deductive structure of rational-choice models enables one not only to sort out better from worse strategies but also to analyze their sensitivity to different modeling assumptions. Behavior inexplicable by one set of assumptions may be eminently rational according to another set.

The models I build to explicate rational choices, together with the formal calculations embedded in the models, are the heart of this book. Any number of examples could be used to illustrate these calculations. Those presented here, though wide-ranging, are necessarily selective and not meant to be conclusive. In fact, no set of examples could be; the case for rational politics depends on the generality and applicability of the models, not on any specific case studies that exemplify their calculations.

The rational-choice models in this book differ substantially not only in their assumptions—except for rationality, in one form or another—but also in the topics to which they are applied. At various points, for example, I describe the individualistic calculations of players in games with radically different settings; delineate what coalitions of players are likely to form and remain stable in electoral politics and international relations; analyze the properties of voting systems used in small committees, legislatures, large conventions, and international organizations; and characterize different situations vulnerable to the exercise of power. Starting from the assumption that actors are rational with respect to some goals in each case, consequences of their rationality are deduced and compared with their actual behavior—insofar as this is possible—to test the empirical validity of the different models.

Such tests help one to determine whether there is a rhyme and reason to political choices beyond the peculiar circumstances of the moment. In the tens of thousands of public elections held every year in the United States, for example, it is reasonable to suppose that candidates face common and recurrent strategic decisions for which rational-choice models may provide both general insights and verifiable hypotheses.

Of course, such models need to be generalized further, and logical connections among them must be tightened or made more parsimonious. They also have to be applied to new empirical situations, which will suggest revisions that make them less arbitrary and hence more germane to other situations. My larger aim here, however, is not to parade before the reader a series of models and their applications but rather to show the merits of an overall conception of politics that takes as a starting point that political choices are made by thoughtful (and usually intelligent) decision makers who want to realize certain ends. They may be impeded in making rational choices by various environmental constraints, but this does not impugn their rationality as such, only their ability to exercise it.

In fact, certain problems of social choice are ineradicable, as will be seen, and hence pose difficulties that no rational calculations can surmount. From the normative perspective of the reformer, the study of rational politics pinpoints where these difficulties lie and, consequently, what possible trade-offs can be made to ameliorate them.

Although there certainly may be differences about the best means to achieve goals, the most wrenching conflicts in politics usually arise because of disagreements in the goals themselves. Yet even if the goals of actors are antithetical, understanding this fact may not only facilitate understanding their behavior but also clarify differences and identify where (if anywhere) reconciliation may be possible. Thereby rational-choice models may be used for normative purposes—if not to eliminate conflict, then at least to explain why it persists and how, possibly, it may be abated.

Rational-choice models offer not only a calculus for understanding political behavior but also a basis for changing it. Whereas developing and testing models may help explain political phenomena and processes, political reformers seek more than scientific explanation. They wish to change the system by eliminating aspects they consider inimical or unethical.

Reformers will be aided in their task if they better understand the source of the problems they perceive and can model the probable effects of the reforms they propose. In fact, the scientific and the normative study of politics often complement each other. Reformers become more persuasive if their assessments are grounded in rigorous analysis and hard evidence; scientists, when stimulated by reformers, are encouraged to go beyond the here and now and to consider the possible as well as the actual.

Ideally, the study of rational politics will encourage a useful dialogue, even if the goals of scientists and reformers—not to mention the practitioners of politics—differ. In fact, one can be both a scientist and a reformer: after having completed a scientific analysis, a scientist who concludes that reforms are in order and feasible can switch hats and become an advocate. In short, the study of rational politics fosters both good science and good reform, which, after all, are two sides of the same analytic coin.

Political Intrigue in the Bible: Esther 2

2.1 Introduction

It may seem odd indeed to go back to the Bible for examples of rational politics. But, on reflection, there is no good reason why the beginning of Western recorded history should not contain instances of political calculation. In fact, one can argue, the cast of characters in most of the great biblical narratives—God included—generally thought carefully about the goals and consequences of their actions.

To be sure, the Bible is a sacred document to millions of people; it expresses supernatural elements of faith that do not admit of any natural explanations. At the same time, however, some of the great narratives in the Bible do appear to be plausible reconstructions of real events.

I have chosen for analysis Esther, a book of the Hebrew Bible, or Old Testament, in which God is never mentioned. If the political machinations reported in this book can be analyzed in strictly secular terms, the story nonetheless has religious overtones—especially for Jews, for whom it serves as the basis for the holiday of Purim.[1]

Some elementary game theory will be introduced both to aid the strategic exegesis of Esther and to illustrate the application of game theory to the analysis of political conflict. Although the popular notion of a game focuses on entertainment, in game theory choices are not assumed to be frivolous. Quite the contrary: players in games are assumed to think carefully about their choices and the possible choices of other players. The outcome of a game—whether comic or tragic, fun or serious, fair or unfair—depends on individual choices. Yet because these choices may have ramifications not only for the individuals involved but also for an entire people—as in the case of Esther—they are unmistakably political.

Game theory is a tool ideally suited for penetrating the complex decision-making situations often described in the Bible. Because its application requires the careful unraveling of a tangle of character motivations and their effects, it imposes a discipline on the study of these situations that is usually lacking in more traditional literary-historical-theological analyses of the Bible.

The game theory in this chapter is supplemented by verbal explications that use ideas from game theory but not its formal apparatus. Indeed, in some instances a rote application of game-theoretic tools would be silly; at those times I resort to a more informal analysis.

Esther is divided into three main parts: Vashti's demise, Esther's ascension, and Esther's final triumph. In each case, I begin by summarizing the relevant story and then model it by constructing a simplified representation of the strategic situation that the Bible describes, in the process defining and explaining technical concepts in the context of the part of the story being modeled. This approach, both here and in later chapters in which more contemporary games are analyzed, makes the concepts more real and also imparts a better intuitive understanding of their meaning in the empirical situation studied.

2.2 The Demise of Queen Vashti

In the third year of his reign, Ahasuerus, king of Persia and Media, and his queen, Vashti, gave a series of banquets for the men and women, respectively, of Shushan, the capital city. On the seventh day of the banqueting, when the king was "merry with wine" (Esther 1:10),[2] he ordered his seven eunuchs to

> bring Queen Vashti before the king wearing a royal diadem, to display her beauty to the people and the officials; for she was a beautiful woman. But Queen Vashti refused to come at the king's command conveyed by the eunuchs. The king was greatly incensed, and his fury burned within him. (Esther 1:11-12)

But before doing anything rash, Ahasuerus consulted with his sages, who were versed in law and precedent. He asked them what to do about his queen's disobedience. One of the sages began with a shrewd calculation that touched hardly at all on Ahasuerus's personal situation:

> Queen Vashti has committed an offense not only against Your Majesty but also against all the officials and against all the peoples in all the provinces of King Ahasuerus. For the queen's behavior will make all wives despise their husbands, as they reflect that King Ahasuerus himself ordered Queen Vashti to be brought before him, but she would

> not come. This very day ladies of Persia and Media, who have heard of the queen's behavior, will cite it to all Your Majesty's officials, and there will be no end of scorn and provocation! (Esther 1:16-18)

Following up on this prediction of a breakdown in the social-political order, the sage recommended that a royal edict be issued that

> Vashti shall never enter the presence of King Ahasuerus. And let Your Majesty bestow her royal state upon another who is more worthy than she. Then will the judgment executed by Your Majesty resound throughout your realm, vast though it is; and all wives will treat their husbands with respect, high and low alike. (Esther 1:19-20)

This advice pleased Ahasuerus, so the edict was issued.

After the king's anger waned, however, "he thought of Vashti and what she had done and what had been decreed against her" (Esther 2:1), which suggests his ambivalence. But just when the reader's expectation is raised that Ahasuerus's remembrance of Vashti will lead to her being forgiven or even recalled, the king's attendants indicate that the search for a successor has begun. Vashti is never heard from again.

To analyze the choices of Vashti and Ahasuerus initiated by Vashti's refusal to comply with Ahasuerus's order, consider the *outcome matrix* shown in Figure 2.1, which gives the consequences of each player's choices of strategies. Vashti, depicted as the row player, has two strategies in this two-person game: obey (O) or disobey ($\bar{O}$) the king's order. Likewise, Ahasuerus has two choices: depose (D) or don't depose ($\bar{D}$) Vashti.

These are the *strategies,* or courses of action, that each player may follow. A *player* is simply an actor or set of actors who can make strategy choices in a *game,* which is an interdependent situation—the outcomes depend on the choices of *all* players. The game played between Vashti

Figure 2.1 Outcome Matrix of Vashti's Disobedience

		Ahasuerus	
		Depose (D)	Don't depose ($\bar{D}$)
Vashti	Obey (O)	Dismissal without reason (1,1)	Male prerogatives reserved (2,4)
	Disobey ($\bar{O}$)	Disobedience punished (3,3)	Disobedience not punished (4,2)

Key: (x,y) = (Vashti, Ahasuerus)
4 = best; 3 = next best; 2 = next worst; 1 = worst

and Ahasuerus is defined in Figure 2.1 by the strategies of the players and the outcomes to which they lead.

A game in which each player knows the other player's (or players') preferences and rules of play is called a *game of complete information.* The consequences arising from the strategy choices of both players are summarized in Figure 2.1. (The pairs of numbers associated with the different outcomes define the preferences of the players and will be explained momentarily.) Although all possible outcomes appear in Figure 2.1, Ahasuerus did not in fact have the choice of deposing or not deposing Vashti. His move occurred after Vashti's, in full knowledge of her prior choice (O or $\bar{O}$).

For the purpose of evaluating the four possible outcomes that can occur, however, Figure 2.1 will be used. It is a form that will be used in later two-person games to describe the outcomes that can arise—independently of the sequence of moves—when a game tree giving the sequence (illustrated later in this section) is omitted.

In my evaluation, I attempt only to rank the outcomes for each player from best to worst, without attaching any specific *utilities,* or numerical values (for example, measured in terms of money), to these ranks. In the representation shown in Figures 2.1, "4" is considered a player's best outcome; "3," next best; "2," next worst; and "1," worst. Hence, the higher the number, the better the outcome.

The first number, x, in each pair is assumed to be Vashti's preference ranking (row player in Figure 2.1); the second number, y, Ahasuerus's ranking (column player in Figure 2.1). I shall refer to these as *payoffs* to the player, though they constitute only ordinal ranks (that is, orderings of outcomes) and not utilities associated with these ranks. Thus, for example, the payoff (2,4) means the next-worst outcome for Vashti, the best for Ahasuerus.

Notice in this game that what is best (4) for one player is not worst (1) for the other; and what is next best (3) for one player is not next worst (2) for the other. If such had been the case, this would be a game in which the preferences of the players are diametrically opposed, called a *game of total conflict.* Instead, the Figure 2.1 game is a *game of partial conflict,* one in which preferences are neither diametrically opposed nor fully coincidental. It is true that both players rank two outcomes [(1,1) and (3,3)] the same, but the other two outcomes [(2,4) and (4,2)] are partially, though not totally, conflictual.

What is the basis of this evaluation? Consider first the ordering of outcomes from Vashti's perspective. Although the Bible does not say so, it is reasonable to suppose that Vashti did not delight in her role as queen. Subservient to a king she probably did not particularly like or respect, and perhaps even detested, she decided to rebel.

Although rebellion might seem ill-advised, Vashti could have reasoned as follows: if Ahasuerus truly loved her, he probably would depose her as queen but not kill her. Accordingly, the worst outcome Vashti entertains by her defiance is ranked 3; while no longer a queen, she would at least not be at Ahasuerus's beck and call. Between her two worst outcomes associated with obedience, being deposed is clearly a worse fate (1) than not being deposed (2). On the other hand, should Ahasuerus be willing to forget or to gloss over her disobedience, Vashti would have established a healthy precedent that would better enable her to resist future encroachments on her personal freedom. Since she would still be queen, but now with greater independence, her successful defiance of Ahasuerus can be ranked as her best outcome (4).

How does Ahasuerus see the game? Because he cares very much about Vashti, even after being spurned, he does not take lightly to dismissing her. That is why he consults with his sages after his anger has subsided. Of course, Ahasuerus would most prefer that Vashti obey him (4), thereby preserving male prerogatives in the kingdom and making her dismissal unnecessary. But if she does not, he could better tolerate the situation in which she is deposed but alive (3) than that in which she remains as a defiant and unloving queen (2). Clearly, Ahasuerus's worst outcome (1) would be to dismiss her without reason (if she had remained obedient).

Ahasuerus is obviously saddened by his decision to depose Vashti. But he is genuinely persuaded, I believe, that the advice he received from his sage was politically astute, and he is willing to stick by his decree. The king has contradictory feelings—his attraction for Vashti on the one hand, and his political responsibility as king on the other. But the second motif comes to dominate the first: it is time for Ahasuerus to shift gears in his life; therefore, the search for a successor begins.

Does game theory explain the choices of Ahasuerus and Vashti in this biblical game? As noted earlier, the game shown in Figure 2.1, in which the players' strategy choices are assumed to be simultaneous—or, equivalently, made independently of each other—does not depict the game that was actually played. This form was used mainly to describe the four different outcomes that can arise and the preferences of the two players for each of them.

The *sequence* of moves described in the story can be represented by the *game tree* shown in Figure 2.2 (read from top to bottom): Vashti first chooses to obey or disobey Ahasuerus; only then does Ahasuerus choose to depose or not depose her. The facts that Vashti's move precedes Ahasuerus's and that Ahasuerus is aware of Vashti's prior choice mean the game cannot be properly represented as a 2 × 2 game (two players, each with two strategy choices), which is the representation given by the

Figure 2.2 Game Tree of Vashti's Disobedience

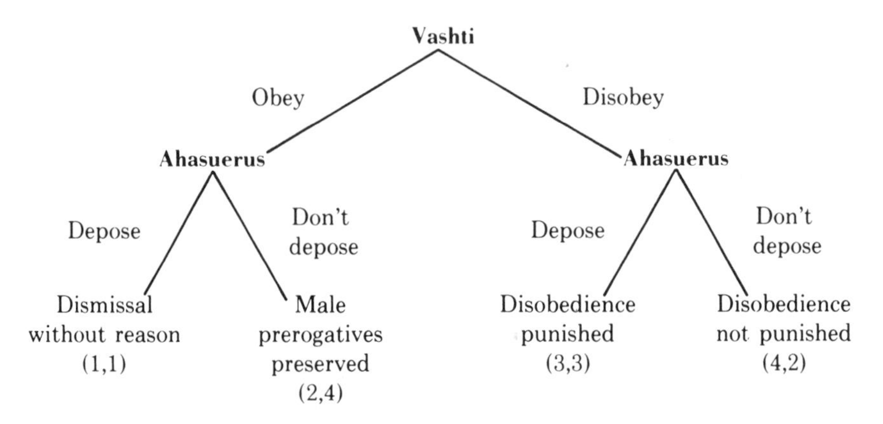

Key: (x,y) = (Vashti, Ahasuerus)
4 = best; 3 = next best; 2 = next worst; 1 = worst

outcome matrix in Figure 2.1. Rather, the proper representation in matrix form of this game is as a 2 x 4 game (Vashti has two strategies, Ahasuerus has four).

This representation is shown in Figure 2.3. The 2 x 4 game reflects the fact that since Vashti has the first move, she can choose whether to obey or disobey Ahasuerus. Ahasuerus, on the other hand, whose moves occur only after Vashti has made a choice, has four possible choices, depending on what Vashti has chosen. Thus, Ahasuerus has four strategies, contingent upon Vashti's prior choices:

1. *D/D depose regardless*. Depose if Vashti is obedient, depose if not.
2. *$\overline{D}/\overline{D}$ don't depose regardless*. Don't depose if Vashti is obedient, don't depose if not.
3. *D/$\overline{D}$ tit-for-tat*. Depose if Vashti is obedient, don't depose if not.
4. *$\overline{D}$/D tat-for-tit*. Don't depose if Vashti is obedient, depose if not.

The 2 x 4 payoff matrix in Figure 2.3 gives the payoffs each player receives for every pair of strategy choices (two for Vashti, four for Ahasuerus) of the two players. Thus, for example, if Vashti chooses to disobey ($\overline{O}$) and Ahasuerus chooses tat-for-tit ($\overline{D}$/D), $\overline{O}$-D is the resultant outcome, for the choice of $\overline{O}$ by Vashti implies the choice of D by Ahasuerus under tat-for-tit. As can be seen from Figure 2.1, this yields a payoff of (3,3)—the next-best outcome for each player—which is shown in the $\overline{O}$ row and $\overline{D}$/D column of Figure 2.3.[3]

Figure 2.3 Payoff Matrix of Vashti's Disobedience

		Ahasuerus			
		D/D	$\bar{D}/\bar{D}$	$D/\bar{D}$	$\bar{D}/D$
Vashti	O	(1,1)	(2,4)	(1,1)	(2,4)
	$\bar{O}$	(3,3)	(4,2)	(4,2)	((3,3)) ← Dominant strategy for Vashti

↑ Dominant strategy for Ahasuerus [column $\bar{D}/D$]

Key: (x,y) = (Vashti, Ahasuerus)
4 = best; 3 = next best; 2 = next worst; 1 = worst
Circled outcome is rational.

What are the game-theoretic implications of the preference assumptions made for Vashti and Ahasuerus? Observe, first, that Ahasuerus's strategy of tat-for-tit ($\bar{D}/D$) is *dominant:* his payoffs associated with this strategy are at least as good as, and sometimes better than, his payoffs associated with any of his other three strategies, whatever Vashti chooses. For example, if Ahasuerus chooses don't depose regardless ($\bar{D}/\bar{D}$), he obtains his best outcome (4) if Vashti is obedient (O), the same as that which he obtains with his dominant strategy ($\bar{D}/D$). On the other hand, if Vashti disobeys ($\bar{O}$), $\bar{D}/D$ yields Ahasuerus a better outcome (3) than does $\bar{D}/\bar{D}$ (2). Similarly with his other two strategies, D/D and $D/\bar{D}$: they never offer better, and sometimes give worse, outcomes than does $\bar{D}/D$.

Thus, $\bar{D}/D$ is Ahasuerus's *unconditionally* best strategy—not dependent on which strategy (O or $\bar{O}$) Vashti chooses—and presumably the choice that a rational player would make in this game. In fact, define for now a *rational player* to be one who chooses a dominant strategy if there is one.[4]

Vashti also has a dominant strategy: disobey ($\bar{O}$) leads to a better outcome than obey (O), whichever of Ahasuerus's four strategies he chooses. In fact her strategy is *strictly dominant,* for there are no ties: this strategy is unequivocally better, whatever Ahasuerus's choice. (Later I shall give examples of games in which at least one player does not have a dominant strategy and discuss what constitute rational choices in such games.)

The choices of $\bar{O}$ by Vashti and $\bar{D}/D$ by Ahasuerus result in outcome (3,3), the next-best outcome for both players, and, in a sense, a kind of compromise outcome in this game. Since this is the outcome that would be chosen by rational players, it is defined to be the *rational outcome.*

This outcome is also *in equilibrium:* once it has been chosen, either player's unilateral deviation from the strategy associated with this outcome would lead to a worse outcome than 3. Because it is also the outcome that the Bible reports was actually chosen by the two players, game theory seems to work well in offering a rational-choice justification for selecting it.

This selection can be justified in another way, based on the game-tree representation in Figure 2.2. The reasoning process is backward, starting near the bottom of the game tree. One asks, first, what the last-moving player (Ahasuerus) would choose if play got to his choice point (or *move*) either on the left—Vashti obeys—or on the right—Vashti disobeys. In the case of obedience, Ahasuerus would choose (2,4) over (1,1); in the case of disobedience, (3,3) over (4,2). Since Vashti, anticipating these subsequent rational choices by Ahasuerus in a game of complete information, would prefer (3,3) over (2,4), it would be rational for her to disobey initially. Then Ahasuerus would depose her, yielding (3,3). This reasoning by *backward induction* leads to the same rational outcome as in the Figure 2.3 game—as indeed it should, since the underlying logic of rational choices in the two different game representations is the same.

There may well be other plausible explanations of the players' choices in this game. For example, in modeling the game, one could make different assumptions about who the players were, what strategy choices were available to them, the outcomes they saw as possible, or their preferences for such outcomes. The reader who is unhappy with the justification I have offered may experiment with other assumptions and test their consequences. Nonetheless, the game-theoretic framework should not be rejected out of hand simply because there are alternative—if not superior—strategic representations of the situations presented here. The main advantage of such a framework is that it forces the analyst to be explicit about assumptions, thereby helping clarify both the basis for differences in the interpretation of strategic situations and the sensitivity of the outcome to these differences.

2.3 Esther's Intercession

After Ahasuerus deposed Vashti, a search was made of the kingdom for a beautiful young virgin to be the new queen. Esther, who had lost both her parents and had been adopted by her uncle, Mordecai, was "shapely and beautiful" (Esther 2:7). When she was brought to King Ahasuerus, he "loved Esther more than all the other women, and she won his grace and favor more than all the virgins. So he set a royal diadem on her head and made her queen instead of Vashti" (Esther 2:17).

Now Esther was a Jew; but on Mordecai's instructions she did not disclose this fact to Ahasuerus. On the other hand, when Mordecai discovered a plot against the king, Esther not only told the king about it—resulting in the execution of the plotters—but also that Mordecai was the source of her intelligence; this fact was duly recorded.

Next to appear on the scene is Haman, who had been advanced by the king to the highest position in the royal court. All bowed down and did obeisance to him except Mordecai. Incensed, Haman "disdained to lay hands on Mordecai alone; having been told who Mordecai's people were, Haman plotted to do away with all the Jews, Mordecai's people, throughout the kingdom of Ahasuerus" (Esther 3:6).

Taking his grievance to Ahasuerus, Haman turned a personal affront into a general indictment of the Jews for insubordination:

> There is a certain people, scattered and dispersed among the other peoples in all the provinces of your realm, whose laws are different from those of any other people and who do not obey the king's laws; and it is not in Your Majesty's interest to tolerate them. If it please Your Majesty, let an edict be drawn for their destruction, and I will pay ten thousand talents of silver to the stewards for deposit in the royal treasury. (Esther 3:8-9)

The king declined the silver but allowed a decree to be issued that all Jews were to be exterminated on a certain day.

There followed great mourning among the Jews. Mordecai, in a message to Esther, asked her to plead with the king on behalf of the Jews. But Esther was fearful of her life, because, as she explained to Mordecai:

> All the king's courtiers and the people of the king's provinces know that, if any person, man or woman, enters the king's presence in the inner court without having been summoned, there is but one law for him—that he be put to death. Only if the king extends the golden scepter to him may he live. Now I have not been summoned to visit the king for the last thirty days. (Esther 4:11)

Mordecai scoffed that Esther should indeed be afraid, for as a Jew she would not escape: "On the contrary, if you keep silent in this crisis, relief and deliverance will come to the Jews from another quarter, while you and your father's house will perish" (Esther 4:14). To this sardonic warning Mordecai added the suggestion that divine providence might have had a role in Esther's elevation to her present royal status: "And who knows, perhaps you have attained to royal position for just such a crisis" (Esther 4:14).

Asking for spiritual assistance from her people, Esther then offered a stoic response:

> Go, assemble all the Jews who live in Shushan, and fast in my behalf; do not eat or drink for three days, night or day. I and my maidens will

> observe the same fast. Then I shall go to the king, though it is contrary to the law, and if I am to perish, I shall perish! (Esther 4:16)

Esther's stoicism is not untouched by logic. Consider the outcome of the game she played with Ahasuerus, as depicted in the matrix in Figure 2.4. Obviously, by interceding before the king, she was playing a risky strategy. But even if she lost and was killed by the king, she could not be faulted for not trying to save her people. Indeed, to them she would be a martyr, so I rate Esther's unsuccessful intercession next best for her (3), her best outcome, of course, being to intercede successfully (4).

Esther's next-worst outcome (2) would be to turn down her uncle, given that Ahasuerus would stop Haman, for then she would be branded a coward for not having tried to save the Jews. Her worst outcome (1) would be to allow the extermination to proceed, for not only would she be disgraced if Haman were not stopped but she also could herself be killed.

Ahasuerus would probably have been untroubled by the mass execution if Esther had not interceded, because he held no special brief for the Jews. Of course, if Esther's religion were made known to him, Ahasuerus would face the dismaying decision of whether to allow her to be executed or make her an exception. But this would not have concerned him when Haman was about to effect his scheme: if Esther did not intercede, he would never be the wiser about the reasons underlying Haman's vendetta against the Jews, much less that his queen was one of them. Accordingly, I rank the status quo, in which Esther does not intercede and Ahasuerus does not countermand Haman's order, to be the king's best outcome (4)—at least before he learns of Esther's identity.

Figure 2.4 Outcome Matrix of Esther's Intercession

		Ahasuerus	
		Stop Haman	Don't stop Haman
Esther	Intercede	Ahasuerus supportive, Esther wins (4,3)	Ahasuerus unsupportive, Esther loses but is martyr (3,1)
	Don't intercede	Ahasuerus inconsistent, Esther cowardly (2,2)	Ahasuerus untroubled, Esther disgraced or killed (1,4)

Key: (x,y) = (Esther, Ahasuerus)
4 = best; 3 = next best; 2 = next worst; 1 = worst

Next best (3) for Ahasuerus would be to support Esther, whom he loved, after she interceded on behalf of the Jews. Next worst (2) would be simply to cancel his order in the absence of Esther's intercession, for then the king would appear inconsistent and weak. Ahasuerus's worst (1) outcome would be not to support Esther after her intercession, because he would lose not only another queen but also the woman he loved. Thus, in a face-to-face showdown with Haman created by Esther's intercession, Esther would win—obtain her best outcome of 4—because 3 is better than 1 for Ahasuerus in the outcome matrix of Figure 2.4.

The problem with this representation—or the 2 x 4 representation of the payoff matrix in which Esther acts first (not shown)—is that if the king refuses to extend his golden scepter to Esther to grant her an audience, the game ends for her. Thus, it is somewhat inaccurate to give Esther the choice of interceding. Perhaps that is why Esther asked her fellow Jews to fast for her. She did not doubt that once she gained an audience with the king she would be able to persuade him of the rightness of her cause. She did worry, though, that he might not deign to see her—that he might even kill her for acting impudently.

I also think Esther worried that if she did not act, the Jews would be saved without her help, as Mordecai predicted, in which case her procrastination would look self-serving and craven. Should this be the case, then effectively the second column of outcomes in Figure 2.4—associated with Ahasuerus's not stopping Haman—can be deleted from the outcome matrix. That is to say, Esther had good reason to believe that Haman's plot would not succeed, even if she kept silent. Given a choice between the payoffs of 4 and 2 in the first column, Esther would obviously choose her strategy associated with the 4 (intercede), thereby reaping the benefit of acting bravely and saving her people and herself.

But this calculation, like that alluded to earlier in which Esther anticipated the outcome of a face-to-face showdown with Haman, does not fully capture the complexity of Esther's choice. After 30 days without seeing Ahasuerus, she could not be at all sure the king would grant her an audience.

2.4 Vindication

That Esther overcame her apprehension in the end attests as much to her astute calculations as to her character. Indeed, though her assessment of the situation was correct, Esther was still very careful to be discreet about implementing her strategy:

> On the third day, Esther put on royal apparel and stood in the inner court of the king's palace, facing the king's palace, while the king was sitting on his royal throne in the throne room facing the entrance of the

> palace. As soon as the king saw Queen Esther standing in the court, she won his favor. The king extended to Esther the golden scepter which he had in his hand, and Esther approached and touched the tip of the scepter. "What troubles you, Queen Esther?" the king asked. "And what is your request? Even to half the kingdom, it shall be granted you." "If it please Your Majesty," Esther replied, "let Your Majesty and Haman come today to the feast that I have prepared for him." (Esther 5:1-4)

At that day's feast, Esther again put the king off, inviting him to come with Haman to a feast she would prepare the next day. Thus, the suspense builds as Esther lays the foundation for Haman's humiliation with drink and charm. Haman was pleased with the first feast, until he saw Mordecai at the palace gate; again Mordecai refused to acknowledge Haman in any way.

Despite his rage at Mordecai's perceived rudeness and arrogance, Haman bragged to his wife of his invitation to Queen Esther's next feast. Haman's wife and friends advised him to erect a stake 50 cubits high and to recommend to the king that Mordecai be impaled on it the day of the banquet.

Then, perhaps by divine intervention, a remarkable coincidence occurred. The king could not sleep, so he ordered that the book of records be read to him; thereupon he learned that Mordecai had never been honored for saving his life from those who plotted against him.

When Haman entered the king's court the next day, the king asked: " 'What should be done for a man whom the king desires to honor?' Haman said to himself, 'Whom would the king desire to honor more than me?' " (Esther 6:6). Haman told the king that this man should be regally attired, with a royal diadem set on his head, and led through the city square on a horse the king had ridden; a proclamation should be made: " 'This is what is done for the man whom the king desires to honor!' " (Esther 6:9). The king then said to a stunned Haman: " 'Get the garb and the horse, as you have said, and do this to Mordecai the Jew, who sits in the king's gate. Omit nothing of all you have proposed' " (Esther 6:10).

The end quickly approached for Haman. At the second banquet the king repeated his question to Esther about her wishes. Now she was finally ready to drop the bombshell:

> "If Your Majesty will do me the favor, and if it pleases Your Majesty, let my life be granted as my wish, and my people at my request. For we have been sold, my people and I, to be destroyed, massacred, and exterminated. Had we only been sold as bondmen and bondwomen, I would have been silent; for the adversary is not worthy of the king's trouble."
>
> Thereupon King Ahasuerus demanded of Queen Esther, "Who is he and where is he who dared do this?" "The adversary and enemy," replied Esther, "is this evil Haman!" (Esther 7:3-6)

Realizing that Haman's charges against the Jews were baseless and that Haman had constructed an odious plot against them, the king then left the wine feast in a fury, leaving a terrified Haman to grovel for his life before Esther. When the king stalked back in, Haman had clumsily undercut his already precarious position:

> Haman was lying prostrate on the couch on which Esther reclined. "Does he mean," cried the king, "to ravish the queen in my own palace?" No sooner did these words leave the king's lips than Haman's face was covered [blanched]. Then Harbonah, one of the eunuchs in attendance on the king said, "What is more, a stake is standing at Haman's house, fifty cubits high, which Haman made for Mordecai—the man whose words saved the king." "Impale him on it!" the king ordered. So they impaled Haman on the stake which he had put up for Mordecai, and the king's fury abated. (Esther 7:8-10)

For anyone with a sense of irony, this kind of biblical justice—being hoist by one's own petard—is hard to rival. But as so often happens in the Bible, the matter does not end with this happy twist for the Jews. Although Mordecai "was now powerful in the royal palace" (Esther 9:4), the Jews still wreaked vengeance on their enemies. Included in the slaughter were Haman's 10 sons, who were impaled on the stake that Haman had built.

The thick plot and heavy irony of this story are almost too astonishing to believe—and hence to model as a serious game played by rational players. I can offer no game-theoretic explanation, for example, why Mordecai first went unrewarded for his good deed to the king, only to be remembered on the day that Haman had slated for his execution. Nor does there seem to be any underlying rationale for Mordecai's instructing Esther to withhold from Ahasuerus her identity as a Jew when she became his new queen. To be sure, the plot depends on Ahasuerus's not knowing the plight of Esther until the end, but the reason for Mordecai's secretiveness in the beginning is somewhat mysterious.

The unexplained coincidences and adventitious events in the Esther story may, of course, result from undescribed divine intervention. What is not beyond natural explanation is Esther's decision to approach Ahasuerus. Recall that Esther shrewdly waited in the wings and hoped that Ahasuerus would see her and beckon her to come before him. This seems to have been a very calculated move on her part to ensure that the king would not be offended by her entry into his presence. Only after her demure entry did Esther consider when intercession would be appropriate and how to prepare for it.

The king seemed enchanted by both Esther and her delicate approach, in marked contrast to the defiant and brash Vashti. After receiving Esther, Ahasuerus immediately promised her up to half his kingdom, but Esther kept gently putting him off. By getting Ahasuerus to repeat his offer several times, it became almost beyond retraction when Esther finally identified the villain she wanted dispatched.

In this matter, Esther masterfully worked her feminine charm. She knew the king's preference for a submissive queen, and she played her role to the hilt. At the core, though, I believe Esther was cunning and hard as nails; the king, it appears, appreciated these qualities, too.

After she had plied Ahasuerus with sweet talk and wine, Esther could hardly have failed in her intercession. Indeed, it is not inconceivable that Esther herself might have set Haman up for the king's return at the second banquet by allowing—perhaps even encouraging—him to assume a compromising position on the couch next to her.

Ahasuerus's adoring queen, then, was also a wily game player who planned her moves with great deftness. Unlike the tactless Mordecai, who insisted on continually flouting Haman, Esther demonstrated how tact could carry the day by smoothing the way for her intercession on behalf of herself and her fellow Jews. Thereby Esther asserted her independence of her uncle while adhering to his basic precepts and following his advice on intercession.

2.5 Politics in Esther

Clearly, the deposition of Vashti did not work out at all badly for Ahasuerus. True, he found it extremely difficult to let Vashti go, pining for her after she was gone. But, for the sake of his kingdom, he knew he could make no other choice than to banish her. Significantly, Ahasuerus decided not to kill Vashti, as she must have sensed would be the case when she made her bid for independence. In the end, Ahasuerus got a less brazen but equally beautiful queen who apparently loved him more than Vashti.

Indeed, Esther is the female embodiment of beauty and charm—and of tact and daring as well. The exquisite use she makes of these qualities in a superbly choreographed *pas de deux* with Ahasuerus, and *pas de trois* with Haman as well, leaves Haman gasping. The irony of Ahasuerus's administering the fatal blow against his favorite courtier makes the plot seem even more fabulous. But, as I have tried to show, the underlying rationale for Esther's actions is neither hard to grasp nor incredible.

But can Esther and Vashti both be considered truly political actors? After all, their conflicts had familial origins, Vashti's with her husband and Esther's with her uncle. Yet each conflict transcended the family and ultimately impinged on many others, giving each conflict a manifest political character, too. In Vashti's case, the political order—not just male prerogatives—would have been threatened, it seems, if other wives had followed Vashti's example and not remained subservient to their husbands. In Esther's case, her personal struggle to maintain her own position required that she broaden the conflict and fight for all Jews in the kingdom.

In his conflict with Vashti, Ahasuerus rebounded quite well from his personal grief, and there was no clear victim. By contrast, Haman was the indubitable victim, along with his family and other allies, in his conflict with Mordecai and Esther. These two cases illustrate that politics, at least in the Bible, has elements of both total conflict (Mordecai and Esther versus Haman, with clear-cut winners and losers) and partial conflict (Vashti and Ahasuerus, in which both players realize a compromise outcome). The latter situation tends to be the norm in politics, but games of total conflict are by no means unheard of in biblical times or later.

2.6 Ethics in Esther

When considering Haman's intention to destroy the Jews, one can unreservedly condemn the actions he took to accomplish this end if one regards the end itself as evil. On the other hand, if one believes Haman had good reason to avenge Mordecai's insubordination, one may see Haman's actions as justified. Yet even granting that Haman was justified in seeking redress for Mordecai's effrontery, the means Haman sought for redress—slaughtering a whole people—could certainly be deplored for being disproportionate to the alleged crime Mordecai committed.

In this biblical tale, therefore, there are at least two ethical issues relating to Haman: the ethics of Haman's goals and of his means for accomplishing them. If either Haman's goals or his means seem detrimental to the cause of justice or plain fairness, it would be proper to reproach his behavior as reprehensible.

Haman first escalated the conflict from a private quarrel with Mordecai to revenge against all the Jews. Esther countered by brilliantly displacing the issue of the Jews' alleged insubordination by the issue of Haman's apparently insidious designs on both herself (when she manipulated Haman into an awkward position on the couch) and Mordecai (who was revealed to be Ahasuerus's savior from an earlier plot on his life). This is not to say that Haman acted stupidly in trying to politicize his private grievance against Mordecai, but rather that he encountered a formidable opponent in Esther.[5] Though his genocidal plot backfired, his depredations against the Jews seemed well thought out as the best way to eliminate Mordecai. That he almost succeeded demonstrates that his scheme was by no means a fantastic one.

Now consider the case of Esther. Although few might question her goal of stopping the wicked Haman from carrying out his dastardly scheme, she was less than forthright in her method, hiding her intent from Ahasuerus until the last moment. Of course, her actions in pursuit of a noble end might be considered just good strategic planning, though they did involve some deception on her part.

What is more questionable, perhaps, is that Esther needed a strong push from her uncle to get her to make an overture to Ahasuerus. Although she agreed to intercede on behalf of her people in the end, she did so because Mordecai reminded her how fragile her own position would be if the position of all Jews in the kingdom should crumble. True, once Esther aligned her goals with the desires of all Jews, she achieved these goals with a vengeance. Not only was Haman dispatched but his entire family was also wiped out—an action that certainly would not be viewed as morally acceptable today. Nevertheless, Esther's rational actions had what most people would probably consider an awesomely beneficial effect in preventing genocide.

One wonders, however, whether any larger purpose roused Esther to act. She did ask for the prayers of her people, but only after she realized that her fate was inextricably bound to theirs.

The story of Esther illustrates that actions one might regard as both rational and morally incontestable may in fact be the product of calculations of private gain alone. Esther is viewed as a heroine for her actions, not her goals, which actually seem quite selfish. That Esther's purposes were enlarged only after it became clear to her that it was rational to include all Jews in her calculations perhaps can be forgiven because, in the end, she made what most would view as the morally correct decision that prevented the monstrous act of genocide.

NOTES

1. The political intrigue of Esther predates the manipulative politics, or "heresthetics," surveyed beginning with ancient Greece in William H. Riker, "Political Theory and the Art of Heresthetics," in *Political Science: The State of the Discipline,* ed. Ada W. Finifter (Washington, D.C.: American Political Science Association, 1983), pp. 47-67. A more comprehensive analysis, based on a number of case studies, is given in William H. Riker, *Political Manipulation: The Art of Heresthetics* (New Haven, Conn.: Yale University Press, forthcoming), which echoes some of the themes presented here.
2. All passages are from a recent translation, *The Writings: Kethubim* (Philadel-

phia: Jewish Publication Society of America, 1982). This translation of the Book of Esther was published earlier in *The Five Megilloth and Jonah: A New Translation* (Philadelphia: Jewish Publication Society of America, 1968; 2d rev. ed., 1974).

3. Ahasuerus's tat-for-tit strategy might better be interpreted as tit-for-tat, because it says Ahasuerus will not depose Vashti (that is, will cooperate) if she obeys (also cooperates); otherwise, he will depose her. The fact that "tit" is equated with "depose"—Ahasuerus's noncooperative strategy—is the reason for this switch in the commonsensical meaning of tit-for-tat.
4. It is easy to show that the expansion of every 2 x 2 ordinal game in which preferences are strict (that is, whose four outcomes can be ranked from best to worst without ties) to a 2 x 4 ordinal game always results in the second-moving player's having a dominant strategy, whether he had one or not in the 2 x 2 game. (Ahasuerus did not: in the Figure 2.1 game, D is better for him if Vashti chooses $\bar{O}$, but $\bar{D}$ is better if she chooses O.) In later chapters I shall address challenges that have been made to the rationality of choosing dominant strategies in certain games.
5. An excellent account of the devastating effect the politicization of issues can have on both the nature of conflict and the outcomes likely to occur is given in E. E. Schattschneider, *The Semisovereign People: A Realist's View of Democracy in America* (New York: Holt, Rinehart and Winston, 1960).

Candidate Strategies 3

3.1 Introduction

The jump of perhaps 2,500 years from Esther's time to the present is an enormous one in recorded history, but the wheelings and dealings of political leaders in the two eras are strikingly similar. What has changed radically are the political institutions for channeling conflict, particularly those related to aggregating individual choices into social choices through voting and public elections.

Elections as we know them today did not occur in biblical times. Saul, the first Israelite king, apparently surfaced because he was "an excellent young man; no one among the Israelites was handsomer than he; he was a head taller than any of the people" (1 Samuel 9:2). Previously, when the Israelites were clamoring for a king "like all other nations" (1 Samuel 8:5), God had told Samuel, the first prophet and judge of Israel, that "it is not you that they [the Israelites] have rejected; it is Me they have rejected as their king" (1 Samuel 8:7). Grudgingly, God then instructed Samuel to "heed their demands and appoint a king for them" (1 Samuel 8:22).

Thus was Saul chosen, who was, by his own self-deprecating testimony, "only a Benjamite, from the smallest of the tribes of Israel, and my clan is the least of all the clans of the tribe of Benjamin!" (1 Samuel 9:21). Such modesty would behoove candidates today!

Certain choices by lot are also reported in the Bible, but only later in Greece and Rome were more systematic procedures developed for the selection of political leaders.[1] Regular, government-supervised elections, relatively free of corruption by the contending parties and of fraud by election officials, are now commonplace in virtually all Western democracies and a few other countries as well. Nevertheless, most of the world is still not subject to democratic rule, under which citizens have traditional civil liberties and there are periodic elections in which all candidates have essentially the same opportunity to compete.

In many developing countries and most Communist regimes, there is either one government-sanctioned party or only token opposition. Blatant manipulation of election returns by the army or other groups ready to step in and overturn a party or candidates not to their liking is also common in nondemocratic countries. Elections in such countries will not be analyzed here, where the game theory of the last chapter—and that to be developed later—seems more appropriate for elucidating the jockeying for favorable positions among the ruling elite than the competition among different elites. (For an application of game theory in the former situation, see Section 6.2 on the struggle for internal control in Poland in 1980-1981.)

In this chapter the focus will be on the spatial games candidates play along a single left-right ideological continuum; models will be used to determine optimal positions of candidates and may be generalized to include multiple dimensions, whose analysis will not be taken up here.[2] Although the one-dimensional spatial analysis of campaign strategies in this chapter is intended to be general, obviously there are vast differences between the campaign strategies of candidates running for local school boards and those running for president of the United States. The analysis of this chapter is most applicable to those running for high office, including president, in which a candidate's positions on key issues tend to be more important, and left-right labels more apposite, than in local races. In Chapter 8 I shall introduce the factor of campaign resources and suggest trade-offs candidates may face between maximizing their resources through relatively narrow ideological appeals and maximizing their votes through moderating their positions.

Party primaries often attract many candidates, whereas the contest in general elections is usually between the two major-party candidates. In this chapter, I start with two-candidate races but then examine the effects of additional candidates' entering the field, paying particular attention to the sequential effects of presidential primaries.

3.2 The Primacy of Issues and Their Spatial Representation

It will be assumed here that voters respond to the positions candidates take on issues. This is not to say that other factors, such as personality, ethnicity, religion, and race, have no effect on election outcomes but rather that issues take precedence in a voter's decision. Although these other factors sometimes become issues in their own right, in the subsequent analysis *issues* will be assumed to be questions of public policy—what the government should and should not do on matters that directly or indirectly affect its citizens.

The primacy of issues in presidential elections has been reasonably well documented over the last twenty years.[3] While most of the research that has been conducted applies to the general election, it would seem even more applicable to primaries, in which the candidates' party affiliation is not a factor.[4] To be sure, some candidates in a primary may claim to be the only true representatives of their party's historical record and ideology. But by making this claim, they are not so much invoking their party label to attract votes as saying that their positions on issues more closely resemble the positions of their party forebears than those of their opponents.

How can the positions of candidates on issues be represented? Start by assuming that there is in a campaign a single overriding issue on which all candidates must take a definite stand. (Later in this chapter candidates will be allowed to fuzz their positions—and thereby adopt strategies of ambiguity—and in Chapter 4 it will be assumed that they can take positions on more than one issue.) Assume also that the attitudes of party voters on this issue can be represented along a left-right continuum, which may be interpreted to measure attitudes that range from very liberal (on the left) to very conservative (on the right).[5] It is not important here exactly what "liberal" and "conservative" mean; this terminology is used only to indicate that the attitudes of voters can be scaled along some policy dimension and given an interpretation to which the words "liberal" and "conservative" can in some way be attached.

The positions candidates take on this dimension or issue are assumed to be perceived by voters in the same way—that is, there is no misinformation about where on the continuum each candidate stands. Like all theoretical assumptions used to model empirical phenomena, this assumption simplifies the reality of candidates' positions and voters' perceptions, but it serves as a useful point of departure for the subsequent analysis.

To derive conclusions about the behavior of voters from assumptions about their attitudes and the positions candidates take in a campaign, some assumption is necessary about how voters decide for whom to vote. More important than the attitudes of *individual* voters, however, are the *numbers* who have particular attitudes along some liberal-conservative scale.

For this purpose I postulate a *distribution* of voters, as shown in

Figure 3.1 Two Candidates and Symmetric, Unimodal Distribution

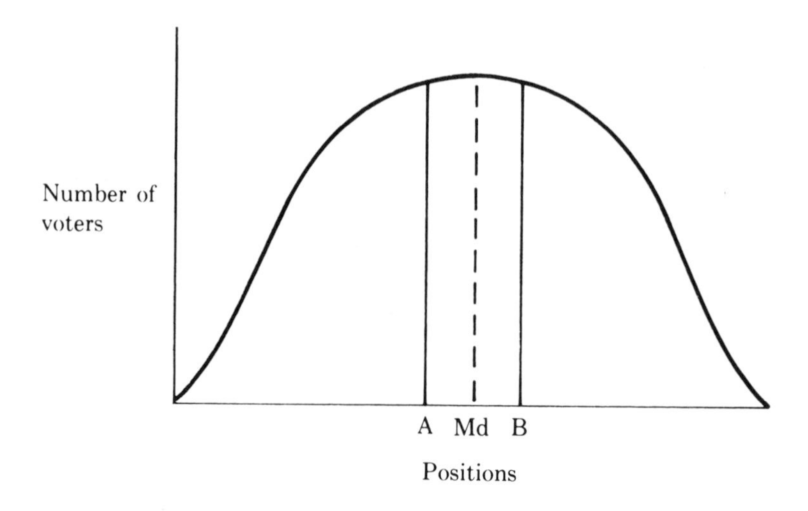

Figure 3.1. The vertical height of this distribution, which is defined by the curve in Figure 3.1, represents the number (or percentage) of voters who have attitudes at each point along the horizontal continuum.[6]

Because the postulated distribution has one peak, or *mode,* it is characterized as *unimodal.* Its *median* is the point where the vertical dashed line intersects the horizontal axis, dividing the area under the distribution curve exactly in half. This means in Figure 3.1 that half the voters have attitudes to the left of the point where the median line intersects the horizontal axis and half the voters have attitudes to the right of this point. Moreover, because the distribution is *symmetric*—the curve to the left of the median is a mirror image of the curve to the right—the same numbers of voters have attitudes equal distances to the left and right of the median.

Voters and candidates, like other political actors, are assumed to have goals and to act rationally to satisfy them in an election. More specifically, assume that voters vote for the candidate whose position is closest to their own along the continuum; assume also that candidates try to choose positions that maximize, in light of the voters' rationality, the total number of votes they receive.[7]

Whereas the *attitudes* of voters are a fixed quantity in the calculations of candidates, the *decisions* of voters will depend on the positions that the candidates take. Assuming the candidates know the distribution of voter attitudes, what positions for them are rational?

3.3 Rational Positions in a Two-Candidate Race

Assume that there are only two candidates in a race, and the distribution of voters is symmetric and unimodal, as illustrated in Figure 3.1. Let candidates A and B take the positions along the left-right continuum shown in Figure 3.1, with candidate A somewhere to the left of the median and candidate B somewhere to the right. Then A will certainly attract all the voters to the left of his or her position, and B all the voters to the right of his or her position. If both candidates are an equal distance from the median, they will split the vote in the middle—the left half going to A and the right half going to B. The race will therefore end in a tie, with half the votes (everything to the left of the median) going to A and half the votes (everything to the right of the median) going to B.

Could either candidate do better by changing his or her position? If B's position remains fixed, A could move alongside B, just to B's left, and capture all the votes to B's left. Since A would have moved to the right of the median, he or she would, by changing position in this manner, receive a majority of the votes and thereby win the election.[8] But, by analogy, there is no rational reason for B to stick to his or her original position to the right of the median. B should approach A's original position to capture all the votes to A's right. In other words, both candidates, acting rationally, should approach each other and the median. Should A move rightward past the median but B move leftward only as far as the median, B would receive not only the 50 percent of the votes on the left but also some of the votes that fall between B's (median) position and A's position (now to B's right). Clearly, A loses by crossing the median. Hence, there is an incentive for both candidates not only to move toward the median but also not to overstep it.

The consequence of these calculations is that the median position is optimal for both candidates. If they both adopted the median position, voters would presumably be indifferent to the choice between the two candidates on the basis of their positions alone and would make their choice on some other grounds.

More formally, the median position is *optimal* for a candidate if there is no other position that can guarantee a better outcome (that is, more votes), regardless of what position the other candidate adopts. Naturally, if B adopted the position shown in Figure 3.1, it would be rational for A, in order to maximize his or her vote total, to move alongside B, as has already been demonstrated. But if B did not remain fixed but instead switched position (say, to the median), the nonmedian position of A would not *ensure* him or her of 50 percent of the votes. Thus, the median is optimal in this example in the sense that it guarantees a candidate at least 50 percent of the total vote no matter what the other candidate does.

The median is also stable in this example because, if one candidate adopts this position, the other candidate has no incentive to choose any other position. More formally, a position is *in equilibrium* if, given that it is chosen by both candidates, neither candidate is motivated unilaterally to depart from it. (See Section 2.2 for the analogous concept in a matrix game.) Thus, the median in the example is both optimal (offers a guarantee of a minimum number of votes) and in equilibrium (once chosen by both candidates, there is no incentive for either unilaterally to depart from it).

A surprising consequence of all two-candidate elections is that *whatever* the distribution of attitudes among the electorate, the median loses none of its appeal in a single-issue election. Consider the distribution of the electorate in Figure 3.2, which is *bimodal* (that is, has two peaks) and is not symmetric. Applying the logic of the previous analysis, it is not difficult to show that the median is once again the optimal, equilibrium position for two candidates.

In this case, however, the *mean* (Mn), which is the "weighted average" position on an issue, does not coincide with the median. As an illustration of this point, consider, as a rough approximation to the continuous distribution in Figure 3.2, the following discrete distribution

Figure 3.2 Nonsymmetric, Bimodal Distribution in Which Median and Mean Do Not Coincide

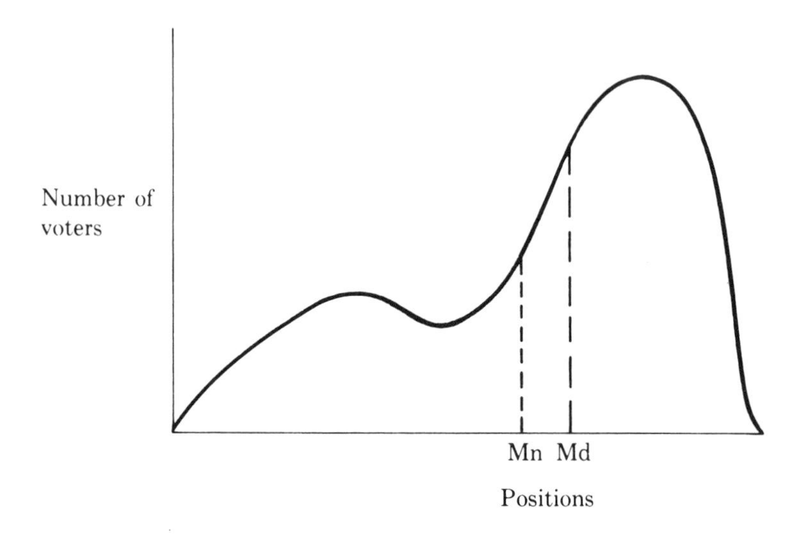

of 19 voters, whose positions on a 0-1 scale are as follows:

1 voter at 0.1
3 voters at 0.2
2 voters at 0.3
2 voters at 0.5
3 voters at 0.6
6 voters at 0.8
2 voters at 0.9

The median of this distribution is 0.6, since 8 voters lie to the left and 8 to the right. The mean, however, is

$$\begin{aligned} \text{Mn} &= (1/19)[1(0.1) + 3(0.2) + 2(0.3) + 2(0.5) \\ &\quad + 3\,(0.6) + 6(0.8) + 2(0.9)] \\ &= (1/19)(10.7) = 0.56 \end{aligned}$$

Taking a position at 0.56 against an opponent who takes a position at 0.6, a candidate would lose 11 votes to 8.

In Figure 3.2 the distribution is skewed to the right, which necessarily pushes the median to the right of the mean. The lesson derived from Figure 3.2 is that it may not be rational for a candidate to take a position at the mean if the distribution of attitudes of the electorate is skewed to the left or right.

A sufficient condition for the median and mean to coincide is that the distribution be symmetric, but this condition is not necessary: the median and mean may still coincide if a distribution is nonsymmetric, as illustrated in Figure 3.3. When the median and mean coincide, a candidate need not take a different position to ensure victory—or at least prevent defeat, if his or her opponent adopts the same position. However, as Figure 3.3 indicates, the noncoincidence of the median and mean is not necessarily related to the lack of symmetry in a distribution: half the voters may still lie to the left and half to the right of the mean (as well as the median) if the distribution is nonsymmetric.

Given the desirability of the median position in a two-candidate, single-issue election, is it any wonder why candidates who strive to win try so hard to avoid extreme positions? As in Figures 3.2 and 3.3, even when the greatest concentration of voters does not lie at the median but instead at a mode (the mode to the right of the median in both these figures), a candidate would be foolish to adopt this modal position. For although the right-leaning voters would be very pleased, the candidate's opponent could win the votes of a majority by sidling up to this position but still staying to the left of the mode.

Voters on the far left may not be particularly pleased to see both candidates situate themselves at the median near the right-hand mode in Figure 3.2, but in a two-person race they have nobody else to whom to

Figure 3.3 Nonsymmetric, Bimodal Distribution in Which Median and Mean Coincide

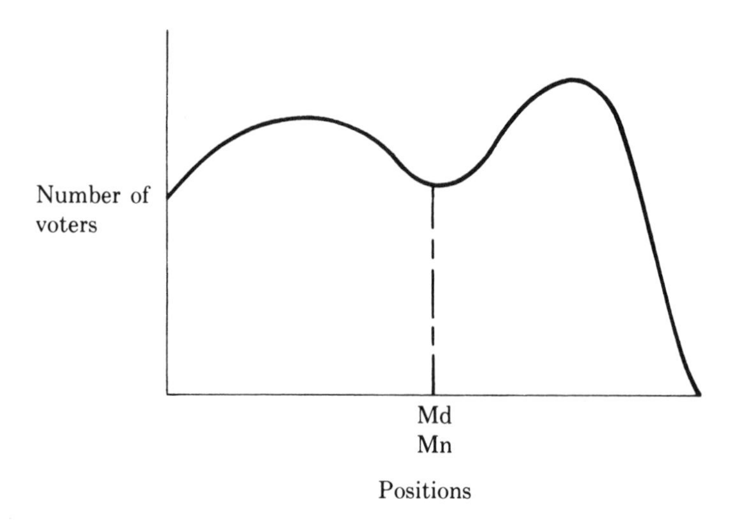

turn. Of course, if left-leaning voters should feel sufficiently alienated by both candidates, they may decide not to vote at all, which has implications for the analysis that will be explored in Section 3.7.

3.4 Rational Positions in a Multicandidate Race

The rationality of entry into a political race is an interesting but almost totally neglected question in the study of elections. Because presidential primaries tend to attract many candidates, especially at the start of the sequence, it is useful to ask what conditions make entry in a multicandidate race attractive.

If no positions that a potential candidate can take in a primary offer any possibility of success, then it will not be rational to enter the race in the first place. For a potential candidate, then, the rationality of entering a race and the rationality of the positions he or she might take once in the race really are two aspects of the same decision.

Assume that two candidates have already entered a primary and, consistent with the analysis in Section 3.3, that they both take the median position (or positions close enough to it to be effectively indistinguishable). Is there any room for a third candidate?

Consider Figure 3.1, but now imagine that A and B have both moved to the median and therefore split the vote since they take the same position. Now if a third candidate, C, enters and takes a position on either side of the median (say, to the right), it is easy to demonstrate that the area under the distribution to C's right may encompass *less than one-third* of the total area under the distribution curve and still enable C to win a plurality of votes.

To show why this is so, in Figure 3.4 I have designated for a position of C to the right of A/B (at the median) the portion of the electorate's votes that A/B on the one hand, and C on the other, would receive. If C's area (shaded) is greater than half of A/B's area (unshaded), C will win more votes than A or B. (Recall that A and B split their portion of the vote since they take the same median position.)

Now C's area includes not only the voters to the right of his or her position but also some voters to the left. More precisely, C will attract voters up to the point midway between his or her position on the horizontal axis and that of A/B: A and B will split the votes to the left of this point; C will win all the votes to the right of this point. Since C picks up some votes to the left of his or her position, less than one-third of the electorate can lie to the right and the candidate can still win a plurality of more than one-third of the total vote.

By similar reasoning, it is possible to show that a fourth candidate,

Figure 3.4 Three Candidates and Symmetric, Unimodal Distribution

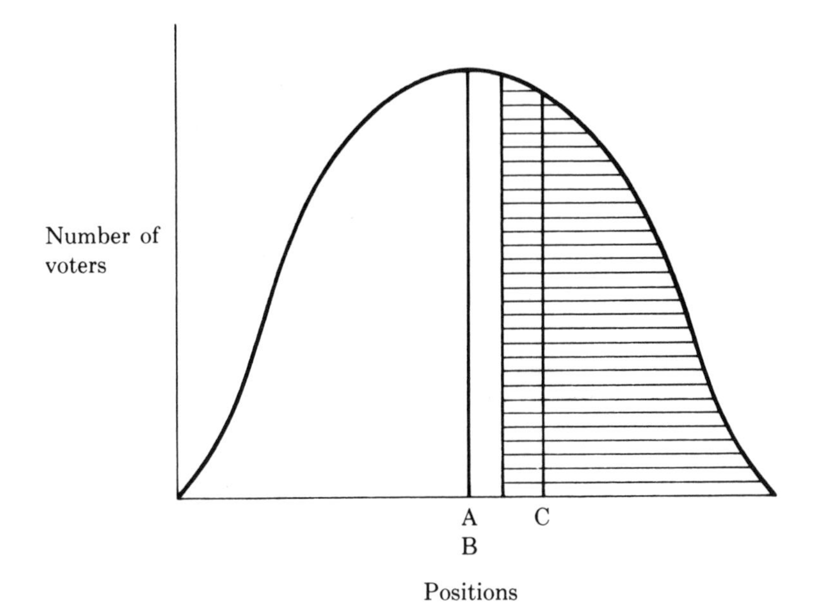

D, could take a position to the left of A/B and further chip away at the total of the two centrists. Indeed, D could beat candidate C as well as A and B by moving closer to A/B from the left than C moved from the right.

Clearly, the median position has little appeal, and is in fact quite vulnerable, to a third or fourth candidate contemplating a run against two centrists. This is one lesson that centrist candidates Hubert Humphrey and Edmund Muskie learned to their dismay in the early Democratic primaries in 1972 when George McGovern and George Wallace mounted challenges from the left and right, respectively. Only after Muskie was eliminated and Wallace was disabled by an assassin and forced to withdraw did Humphrey begin to make gains on McGovern in the later primaries, but not by enough to win.

In fact, there are *no* positions in a race initially involving only two candidates, for practically any distribution of the electorate, in which *at least one* of the two candidates cannot be beaten by a third (or fourth) candidate.[9] I have already shown that *both* candidates in a two-candidate race can be beaten by a third (or fourth) candidate if they both adopt the median position. Indeed, it is easy to show that *whatever* positions two candidates adopt—the same or different—at least one will always be vulnerable to a third candidate; if the other is not vulnerable to the third candidate, he or she will be vulnerable to a fourth.

To understand this, assume that two candidates, perhaps anticipating other entrants and realizing the vulnerability of the median, take different positions, as illustrated in Figure 3.5. In this example, because the distribution is bimodal (as well as symmetric), positions at the modes would seem to be strong positions for each of two candidates to hold.

But enter now a right-leaning third candidate, C, who would like to push candidate B out of the race. Excluding the possibility of ties, either (1) there are more voters to the right of B than between B and the median/mean or (2) the opposite is true. If (1) is true, then C can beat B by moving alongside B to the right; if (2) is true, then C can beat B by moving alongside B to the left. In either event, B is vulnerable to the third candidate, C (and A would be vulnerable to a fourth candidate, D, for similar reasons). Hence, a third (or fourth) candidate can knock out at least one of the two original candidates (A and B) in the example.

The vulnerability to third (or fourth) candidates of *any* positions that two candidates might take in single-issue races can be demonstrated generally, given very nonrestrictive assumptions about the distribution of voter attitudes. There is, in fact, *always* a position a new candidate can choose along a left-right continuum that will displace one or more nearby candidates.

This is not to say, however, that a third candidate will necessarily win against *both* A and B. There are both obstacles and opportunities for

Figure 3.5 Two Candidates and Symmetric, Bimodal Distribution

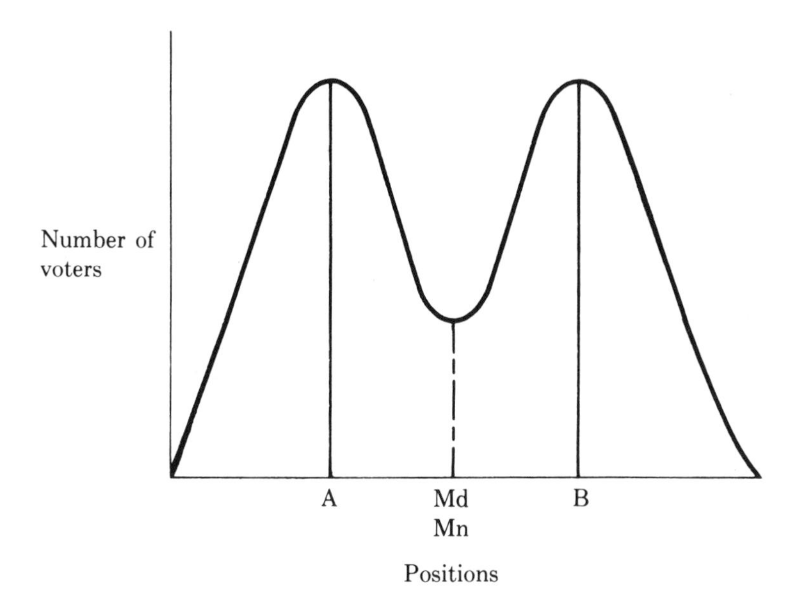

candidate C, which may be summarized by the following propositions:[10]

(1) *1/3-separation obstacle.* If A and B are equidistant from the median of a symmetric distribution and separated from each other by no more than 1/3 of the area under the curve (that is, no more than 1/3 of the voters lie between them), C can take no position that will displace both A and B and enable C to win.

(2) *2/3-separation opportunity.* If A and B are equidistant from the median of a symmetric, unimodal distribution and separated from each other by at least 2/3 of the area under the curve (that is, at least 2/3 of the voters lie between them), C can defeat both A and B by taking a position at the median (exactly between them).

Essentially, if there is little room in the middle, a third candidate can beat one candidate but will then hand the election to the other. This is what occurred in the 1912 presidential election: Theodore Roosevelt, with 27 percent of the popular vote, beat William Howard Taft, with 24 percent, when Roosevelt ran as the Progressive party candidate after losing the Republican party nomination to Taft. But both were defeated handily by the Democratic party candidate, Woodrow Wilson, who got 42 percent of the vote (the fourth candidate, Socialist Eugene V. Debs, got 6 percent).

If there is a wide separation between A and B, by contrast, there may be room in the middle for C [the conditions of (2) ensure that C will get more than 1/3, and A and B less than 1/3, of the vote], but this event has never occurred in a presidential election. In fact, the 1912 election is the only one in which even one major-party candidate has been defeated by a third-party candidate in the popular vote.[11]

The vulnerability of at least one of two candidates to displacement by a third provides an explanation, in terms of the rational choices of both voters and candidates, of why many candidates may initially be drawn into the fray in presidential primaries. Consider these cases in point: in the Democratic party primary in New Hampshire in 1976 (the first in the nation), four candidates each received more than 10 percent of the vote (Jimmy Carter won with 28 percent); four years later four candidates in the Republican party primary each received at least 10 percent of the vote (Ronald Reagan won with 50 percent); in 1984 three candidates in the Democratic party primary surpassed 10 percent (Gary Hart won with 37 percent), with three of the five other candidates each drawing 5 percent or more.

To be sure, if an incumbent president or vice president is running, or even contemplates running, members of his party may be deterred from entering the primaries because of the built-in advantages that his incumbency brings.[12] But it should be pointed out that incumbency did not stop Eugene McCarthy from challenging Lyndon Johnson in the 1968 Democratic party primaries, Paul McCloskey from challenging Richard Nixon in the 1972 Republican party primaries, Ronald Reagan from challenging Gerald Ford in the 1976 Republican party primaries, or Edward Kennedy from challenging Jimmy Carter in the 1980 Democratic party primaries.

Generally speaking, most primary challenges that have been mounted against an incumbent in recent presidential elections have been single-man crusades and can be viewed, therefore, as essentially two-candidate contests. On the other hand, when an incumbent does not run, the field opens up and many candidates are motivated to stake out claims at various points along the left-right continuum.

3.5 The Winnowing-Out Process in Presidential Primaries

To explain the entry of multiple candidates into presidential primaries, I previously considered the contest for the nomination as if it were one election in which each candidate sought to maximize his vote total. But this limited perspective clearly does not explain the *exit* of candidates from presidential primaries. Indeed, probably the most important feature of these primaries distinguishing them from other elections is their sequential nature; it is performance *in a particular sequence*—not in one primary election—that is crucial to a candidate's success.

The importance of sequencing is conveyed quite dramatically by statistics from the 1972 Democratic party primaries. In these primaries, roughly 16 million votes were cast, with George McGovern polling 25.3 percent of the total primary vote and Hubert Humphrey 25.4 percent, despite entering later.[13] Nonetheless, though McGovern received fewer primary votes than Humphrey, and little more than a quarter of the total, he went on to win his party's nomination on the first ballot at the national convention. Bone and Ranney attribute McGovern's success to "certain breaks," but it seems that a winning strategy in a series of primaries is more than a matter of luck.[14] I shall not try to analyze McGovern's success specifically, however, but rather attempt to identify optimal strategies over a sequence of elections generally.

One is immediately struck by the fact that as an institution presidential primaries play a role less in selecting candidates than in eliminating them. Candidates who have won or done well in the primaries, such as Estes Kefauver in the 1952 Democratic party primaries or Eugene McCarthy in the 1968 Democratic party primaries, have, despite their impressive showings, lost their party's nomination to candidates who did not enter the primaries (Adlai Stevenson in 1952, Hubert Humphrey in 1968).

Recently, however, incumbent presidents, including Gerald Ford in 1976 and Jimmy Carter in 1980, have had to fight hard for their party's nomination in presidential primaries. Moreover, when there has been a primary fight, no candidate ever defeated in the primaries has ever gone on to capture his party's nomination in the convention.

Once a candidate enters the primaries, his or her priority is not to be eliminated. In a multicandidate race, this goal most often translates into not being defeated by an opponent or opponents who appeal to the same segment of the party electorate.

To facilitate the subsequent analysis, assume that there are three identifiable segments of the party electorate: liberal, moderate, and conservative. This trichotomization of the electorate may not always be an accurate way of categorizing different positions in multicandidate races, but these labels are commonly used by the media and the public.

A candidate who takes a position on the left-right continuum will fall into one of these three segments. The candidate will be viewed to be in a contest—at least in the first primaries—with only those other candidates who take positions in the same segment as he or she occupies. What is likely to happen if there are at least three candidates contesting the vote in each segment? More specifically, who is likely to beat whom in the first-round battles and survive the cuts of candidates in each segment?

If the distribution of the electorate is symmetric and unimodal, as pictured in Figure 3.1, then the liberal segment will appear as in Figure 3.6, with the median of this segment to the right of the mean. The median will be attractive in a two-candidate liberal contest (for reasons given in Section 3.3), but should a third candidate battle two candidates who take the median position in this segment, then his or her rational strategy would be to move to the right of the median—and toward the center of the overall distribution—where voters are more concentrated in the liberal and adjoining moderate segments.

This movement toward the center may be reinforced by two considerations, one related to the concentration of votes near the center and the other by an anticipation of future possibilities in the race. If voters may become alienated by a candidate whose position is too far from their own and respond by not voting, a candidate can minimize this problem by being to the right rather than the left of the median in Figure 3.6, where a loss of voters more than a given distance from his or her position will be numerically less damaging (as will be discussed in Section 3.7). In addition, a position to the right of the median becomes more attractive as

Figure 3.6 Liberal Segment of Symmetric, Unimodal Distribution

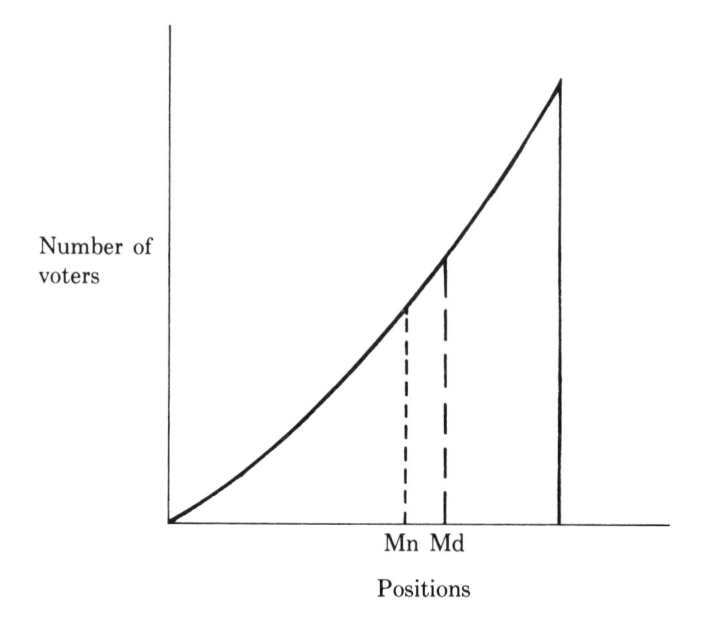

Figure 3.7 Moderate Segment of Symmetric, Unimodal Distribution

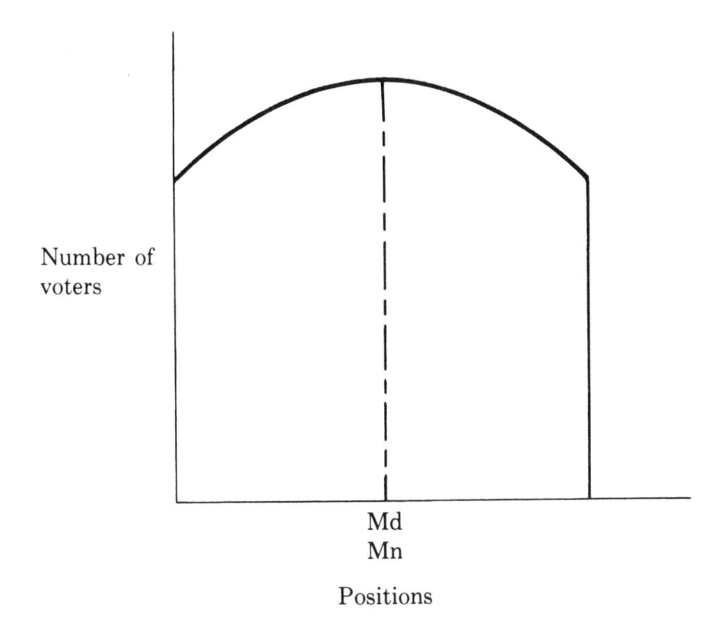

moderate candidates are eliminated and the liberal survivor can begin to encroach on voters who fall into the moderate segment.

Thus, liberal candidates will be motivated to move toward the moderate segment and, for analogous reasons, conservative candidates will also be motivated to move toward the moderate segment (though from the opposite direction). What should the moderates do in their own segment?

If, for the distribution shown in Figure 3.7, two candidates take the median position, which is also the mean because of the symmetry of this segment, then a third moderate candidate would be indifferent between taking a position to the left or right of the median/mean since voters are symmetrically distributed on either side. To illustrate the consequences of a nonmedian position, assume that the third candidate takes a position somewhat to the right in the moderate segment. This candidate thereby captures a plurality of the moderate vote against the two opponents at the median and eliminates them from the contest (see Section 3.4).

As a consequence of these outcomes, the election might well reduce to a three-way contest among a liberal (L), a moderate (M), and a conservative (C), with positions approximately as shown in Figure 3.8. In this manner, the initial primaries serve the purpose of reducing candidates to just one in each segment.

Figure 3.8 Three-Way Contest among Liberal, Moderate, and Conservative Candidates

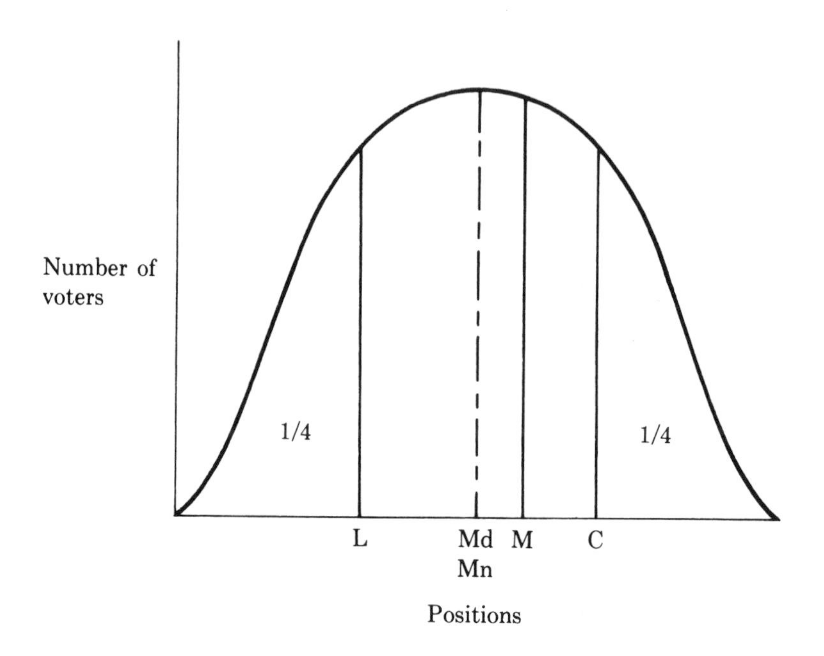

But the elimination process does not stop here. In fact, if as few as 1/4 of the voters lie to the left and 1/4 to the right of the liberal and conservative candidates, respectively (see Figure 3.8), it is unlikely that the moderate candidate will get the most votes. Since in this example the moderate is not at the median but to its right, he or she will in all likelihood receive hardly more than 1/2 of those votes in the middle (or 1/4 of the total, since 1/2 of the total falls between L and C).[15]

Hence, the moderate candidate will probably receive fewer votes than both the liberal and conservative candidates, for these candidates will pick up all the votes to their left and right, respectively (1/4 of the total), plus all the votes in the moderate segment up to the point midway between their positions and that of the moderate candidate. This would obviously help L, who is farther from M than C is. In fact, if L can supplement his or her 1/4 support to the left with somewhat more than 1/12 of the total votes from the moderate segment and C can obtain at least 1/12 in this segment, then L would receive more than 1/3 of the total and C at least 1/3, thereby limiting M to less than 1/3.

This is why a liberal candidate like McGovern could win with slightly more than 25 percent of the primary votes. More generally, a moderate candidate can be squeezed out of the race by challengers on both sides of the continuum even when the bulk of voters falls in the middle. If most voters are not concentrated in the middle, but tend instead to be either liberal or conservative, then of course the problems of a moderate are aggravated.

Even if most voters are concentrated in the middle, the moderate may face another kind of problem. It is possible for more than one moderate to attract a sufficient number of votes to survive the early primaries. But opposed by just one surviving liberal and one surviving conservative in the later primaries, the two or more moderates who divide the centrist vote will lose votes relative to the liberal and conservative candidates who have picked up votes from those eliminated from their segments. The 1964 Republican party primaries are an example of this situation: Henry Cabot Lodge, Jr., a moderate, lost out to Nelson Rockefeller and Barry Goldwater, the liberal and conservative candidates who fought a final climactic battle in the California primary, won by Goldwater.

Moderates are not inevitably displaced in a sequence of primaries—as the case of Jimmy Carter in the 1976 Democratic party primaries demonstrates—but such displacement has been one trend in recent years in heavily contested primaries in both parties. (Ronald Reagan in the 1980 Republican party primaries, for example, beat several more moderate Republicans.) As I have tried to show, spatial analysis enables one to understand quite well the weakness of moderates when squeezed from the left and right in a series of elimination contests.

3.6 The Factor of Timing in Presidential Primaries

Presidential primaries, I have suggested, are first and foremost elimination contests that pare down the field of contenders over time. Implicit in the previous analysis has been the assumption that the key to victory in the primaries is the position that a candidate takes on a left-right continuum *in relation to the positions taken by other candidates.* Thus, a candidate's goal of avoiding elimination, and eventually winning, cannot be pursued independently of the strategies other candidates follow in pursuit of the same goal. This quality of primaries, and elections generally, is what gives such contests the characteristics of a game, in which winning depends on the choices that *all* players make.

Since the rules of primaries do not prescribe that these choices be simultaneous, there would appear to be advantages in choosing *after* the other players have committed themselves and the strengths and weaknesses of their positions can be better assessed.[16] Indeed, some candidates—especially if they are well-known—have avoided the early primaries and joined the fray at a later stage on the basis of just such strategic calculations. Robert Kennedy, for example, stayed out of the 1968 Democratic party primaries until the weakness of Lyndon Johnson's position as the incumbent became apparent; he successfully engaged Eugene McCarthy in the May 7 Indiana primary and most others until his assassination on June 5, after the California primary.

A more extreme case of a late-starter was Hubert Humphrey, who stayed out of the 1968 Democratic party primaries altogether, apparently believing that as the incumbent vice president he stood his best chance in the national party convention. He was not to be disappointed, winning on the first ballot; in the end his only serious opposition came from McCarthy because of the earlier assassination of Kennedy.

The advantages of starting late, when the positions of one's opponents are known and their weaknesses can be identified and exploited, must be balanced against the organizational difficulties one faces in launching a campaign hurriedly. Last-minute efforts, even by well-known candidates, have often fizzled out. On the other hand, candidates who are not household names, like Eugene McCarthy in 1968, George McGovern in 1972, Jimmy Carter in 1976, John Anderson in 1980, and Gary Hart in 1984, have no choice but to start their campaigns very early to acquire sufficient recognition to make a serious run.

How can spatial analysis be used to model the factor of timing? Consider the situation in which several candidates to the left and right of the median struggle for their party's nomination in the early primaries. Assume that their various positions fall within the shaded bands pictured in Figure 3.9, in which the distribution of voter attitudes is assumed to be symmetic and unimodal.

Assume also that a prominent moderate politician considers making a bid for his or her party's nomination by positioning himself or herself somewhere near the median/mean. The candidate calculates that his or her chances of winning the party's nomination are good if extreme (e) candidates are the ones to survive in the early primaries on the left and right (at positions L_e and C_e), since the moderate will be able to capture the bulk of the votes in the middle of the distribution. On the other hand, if other moderate (m) candidates are the ones to survive in the early primaries (at positions L_m and C_m), the new entrant will probably be squeezed out by one or the other if he or she runs.

Thus, to gain a better picture of his or her chances, the prominent moderate may decide to await the results of the early primaries before making a decision, even if it means postponing the building of a campaign organization that would strengthen his or her bid. Yet there may be a more compelling reason to avoid an announcement, based on spatial considerations.

Figure 3.9 Bands Encompassing Positions of Candidates on Left and Right

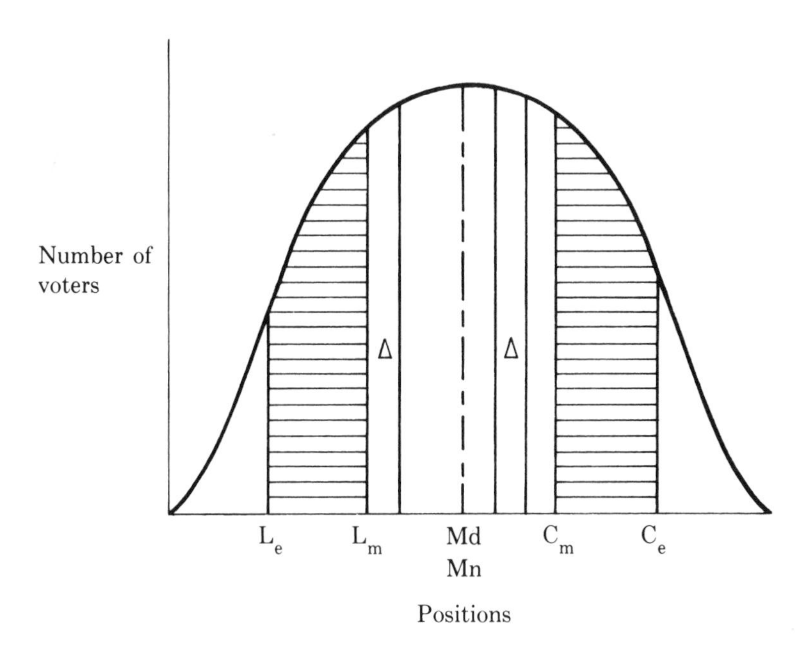

Assume that the survivors of the early primaries are an extreme liberal candidate (at L_e) and a moderate conservative candidate (at C_m). Thus, if the moderate runs, he or she will be squeezed more from the right than from the left. Clearly, the moderate's chances are not so favorable as they would be if the race were against two extreme candidates on the left and right. Nonetheless, spatial analysis clarifies how he or she can capitalize on the information gained from awaiting the results of the early primaries to take an optimal position against his or her surviving opponents at L_e and C_m.

Although one might think initially that a holdout moderate could maximize his or her vote total by taking a position midway between L_e and C_m, a glance at Figure 3.9 will show this to be a poor strategy. Instead, the candidate should take a position to the right of the median/mean near C_m.

The latter strategy follows from the fact that the votes given up to the L_e opponent as the moderate moves to the right of the median/mean are more than compensated for by the votes gained from the C_m opponent as he or she moves toward that position. It can be seen from Figure 3.9

that there are more votes in the △ region just past the midway point between the median/mean and C_m than in the △ region just past the midway point between L_e and the median/mean. Therefore, a moderate gains more votes in the right △ region than he or she loses in the left △ region by moving rightward toward C_m.

It is evident, then, if the distribution of voter attitudes is symmetric and unimodal, a late-starting moderate's best weapon against opponents on his or her left and right is *to move toward his or her more moderate opponent.* This qualitative analysis does not say exactly how far he or she should move, but the quantitative problem of finding an optimal position can easily be solved if the distribution of voter attitudes is known.[17]

The analysis in this section can be extended to different-shaped distributions and can incorporate different assumptions about the positions of committed candidates and the timing of the announcement of an uncommitted candidate. My main purpose, however, has been to introduce with a simple example the factor of timing into the spatial analysis of primaries, not to try to treat this subject exhaustively; the subject deserves much more systematic attention than it has received in the literature.

3.7 Fuzzy Candidate Positions and Voter Alienation

In the preceding section the possibility was considered that there may be several candidates to the left of the median and several candidates to the right whose collective positions can be represented by bands, rather than lines, on the distribution. This same representation can also be used to model the positions of candidates that are *fuzzy*—that is, that cover a *range* on the left-right continuum instead of occurring at a single point.

Fuzzy positions in campaigns are well known and are reflected in such statements as, "I will give careful consideration to . . ." (all positions are open and presumably equally likely), "I am leaning toward . . ." (one position is favored over the others but is not a certain choice), and "I will do this if such-and-such . . ." (choices depend on such-and-such factors). Such ambiguous statements may be interpreted as probability distributions, or lotteries, over specific positions and have been shown to be rational choices under certain circumstances, not only for candidates but for voters as well.[18] To model fuzzy positions, however, I shall not introduce probabilities into the spatial analysis but instead shall analyze some implications of band versus point positions.

First, though, consider why a candidate may not want to adopt a clear-cut position on an issue. Perhaps the principal disadvantage of clarity in a campaign is that while attracting some voters it may alienate others, independently of the positions other candidates take. That is, voters whose positions on the continuum are sufficiently far from the position a candidate takes may feel disaffected enough not to vote at all, even given the fact that this candidate's position is closer to theirs than are those of the other candidates. These voters may be considered *alienated owing to incompatibility*.

Much has been made of the alienated voter in the voting-behavior literature, with many different reasons offered for this alienation.[19] But the fact of alienation—as measured, for example, by the number of citizens who fail to vote—is indisputable. To be sure, some voters fail to vote because of such legal restrictions as residency requirements or bars against felony convictions, but the vast majority of nonvoters in a presidential election—47 percent in the 1980 and 1984 presidential elections—are eligible but choose not to exercise their franchise. In presidential primaries, by comparison, an even greater proportion of eligible voters—often 60 to 70 percent—do not vote, although typically there are more candidates to choose from than in the general election.

For the purposes of spatial analysis, assume that the alienation of a voter is a direct function of his or her distance from the position of the closest candidate. If this distance is sufficiently great, then the voter's alienation overcomes the desire to vote for the closest candidate, and he or she becomes a nonvoter.

The assumption that voters too far from any candidate's position will be alienated may contravene findings from the earlier analysis. For example, alienation tends to undermine the desirability of the median/mean in Figure 3.3 and to enhance the desirability of the two modes in this figure as the optimal positions in a two-candidate race. The reason is that the number of voters alienated a given distance from the median/mean may be more than the numbers alienated the same distance from either mode. The decrease in the number of alienated voters at the modes implies an increase in voter support, making the modal positions more attractive to the candidates.

Thus, a bimodal distribution in which alienation is a factor may induce rational candidates to adopt polarized positions on the left and right of an issue rather than to locate themselves near the median. While advocates of so-called responsible parties (and candidates) that present clear and distinct choices to the voters would view this polarization as salutary, advocates of compromise would not be enamored of the black-and-white choices such polarization entails.

One way a candidate can reduce his or her distance from voters, thereby possibly avoiding the vote-draining effects of alienation, is to fuzz his or her position. Given that voters perceive a candidate's ambiguity as favorable to them, a strategy of ambiguity will increase the broadness of the candidate's appeal.

Figure 3.10 Fuzzy Position of a Candidate

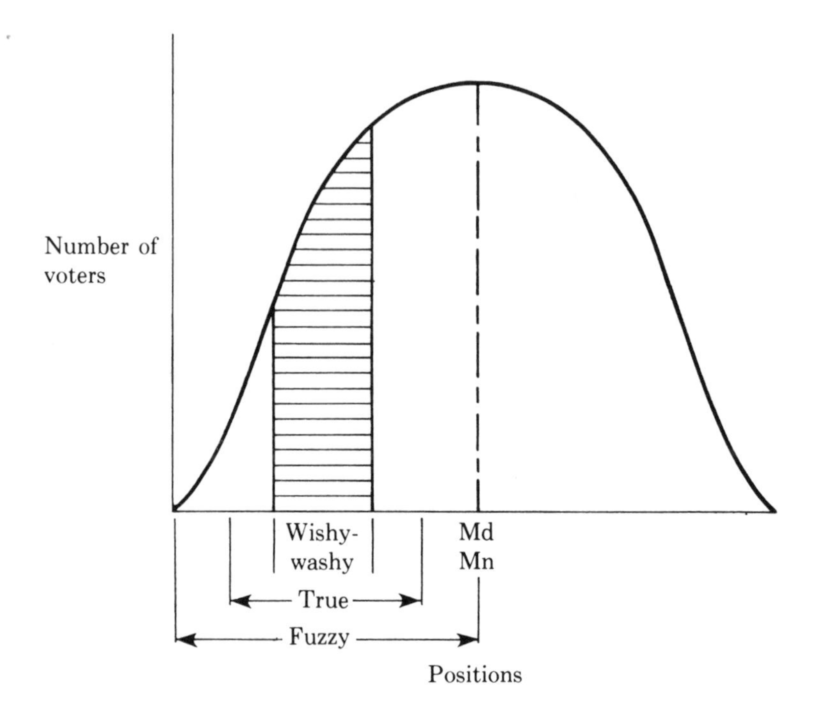

As an illustration of the possible advantages of ambiguity, assume that a candidate's *true position* is at the center of the band in Figure 3.10. If the candidate does *not* fuzz his or her position, assume that the *reach* of this position along the continuum—how far it extends—is that shown as true in Figure 3.10.

If the candidate fuzzes his or her position, however, he or she might be able to extend its reach from the left extreme to the median, assuming that voters on the left extreme interpret the candidate's position to be the left boundary of the band and voters at the median interpret the position to be the right boundary of the band. On the other hand, if voters make the opposite interpretation—that the boundaries of the band *farthest* from them are the actual positions of the candidate—an ambiguous candidate may perversely succeed in contracting (rather than expanding) voters' support by fuzzing his or her true position. Call this candidate's position as interpreted by voters *wishy-washy* and assume its reach to be only the bandwidth itself, which is much narrower than the fuzzy range in Figure 3.10.

Thus, a danger may attend a strategy of ambiguity, depending on what voters perceive to be the actual position of a candidate. Given that they recognize the ambiguous strategy of a candidate to be a band rather than a point on the continuum, their choices may depend on whether they view this ambiguity to represent a desirable flexibility or an undesirable lack of conviction.

Apparently, voters respond to ambiguity differently in different elections. Nobody ever accused Richard Nixon of forthrightness in his 1968 presidential campaign when he said "I have a plan" to end the war in Vietnam. But, judging from the results of the Republican primaries and general election in 1968, more voters believed in his general competence to deal with the Vietnam situation than were persuaded by the more specific proposals of his opponents.

In contrast, voters came to see George McGovern as irresolute in 1972, especially as he became increasingly vague about specific proposals made in the early Democratic primaries in 1972. Then, to make matters worse, he withdrew his initial "1,000 percent" support of his vice-presidential choice, Thomas Eagleton, after the convention, and he replaced Eagleton by Sargent Shriver. At the polls in the general election, voters overwhelmingly chose the known quantity, incumbent Nixon.

Jimmy Carter's positions before and after the 1976 election present an interesting blend in contrasts: during the campaign he was unspecific on a variety of issues, stressing moral and spiritual themes, but after the election he developed numerous detailed programs (for example, on energy and welfare) that he presented to Congress. Ronald Reagan, by contrast, was quite specific in arguing for conservative positions in his 1980 campaign; yet the central issue in that election, other than the poor state of the economy, was probably the question of each candidate's competence to deal with domestic and international problems. The nonissue of presidential leadership became an issue, to Carter's detriment, and he lost the election. Four years later, in 1984, Reagan's perceived decisiveness as president provided Walter Mondale, the Democratic nominee, little opportunity to wrest this issue from Reagan.

These examples indicate that a strategy of ambiguity may be productive or unproductive, depending on how the candidate is viewed by the voters. From a spatial perspective, an ambiguous strategy would seem least risky for a candidate who tries to push his or her support toward the extremes, if the candidate can also hold on to voters whose positions are nearer the center. On the other hand, a candidate squarely but ambiguously in the center is likely to suffer from attacks of opponents on the left and right, which may erode his or her centrist support from both sides, especially if the opponents can represent his or her position to be at the boundary of the band farthest from them.

Admittedly, these conclusions are rather speculative, principally because very little is known about what kinds of factors engender support for or opposition to fuzzy positions. In the absence of such knowledge, one can make only tentative assumptions about the relationship between ambiguous strategies and voting behavior to derive conclusions about optimal positioning.

Although voter alienation is pervasive, its implications are not entirely clear, especially for presidential primaries. To begin with, citizens may fail to vote in the early primaries not so much because they find the candidates unattractive as because they know very little about them. This might be called *indifference due to ignorance:* voters may not even know how to characterize the candidates generally, much less distinguish their specific positions.[20] However, as the field narrows in later primaries and more information is generated about the races in both parties, the positions of candidates, whether specific or ambiguous, become clearer. Then alienation due to incompatibility may begin more and more to manifest itself.[21]

As early contenders are eliminated and the appeal of the surviving candidates broadens, each will feel less need to draw a fine line between himself or herself and the other survivors, who will generally be spaced farther apart along the continuum. Hence, there will be an incentive for a candidate to extend his or her position from a point to a band to take in voters who otherwise would be alienated because they fall between or—if situated at the extremes—too far away from positions that have been eliminated.

On the other hand, the danger of being seen as wishy-washy or evasive, especially in light of attacks from the opposition, may inspire contraction. The frequently observed consequence of buffeting by these contradictory forces is to-and-fro movements as candidates stick close to basic positions while scampering for pockets of support somewhere removed from these positions. It is fascinating to watch this dance performed along the continuum, even if—or perhaps because—it is almost always improvised and sometimes becomes quite frantic.

3.8 Conclusions

The main contention of this chapter is that candidates play games along a left-right continuum, especially in presidential elections. In presidential primaries in particular, the principal goal of a candidate is to avoid elimination long enough to enter the national party convention as a serious contender, if not an outright winner, assuming that winning and not his or her policy position on an issue is what counts most. However, the tables are reversed for the voter, who wants to maximize his or her satisfaction on issues by choosing the candidate with the most compatible positions.

The spatial games candidates play to try to maximize their appeal to voters have been the focus of most of the analysis in this chapter. In some cases, candidates seem to hew to the median, in others to take more extremist positions, but both strategies may be rational depending on the number and positions of the other candidates in the race. Perhaps the most surprising conclusion that emerges from the analysis is that a moderate candidate will maximize voters' support not by taking a position midway between two opponents on the left and right but instead by taking a position closer to his or her less extreme opponent, given a symmetric, unimodal distribution of voter attitudes.

So far the spatial analysis has been restricted to a single dimension in which the positions candidates take on a single issue totally determine the vote they receive. Unlike a general election, however, in which the party affiliation of a candidate may account for a substantial portion of his or her vote independent of the position he or she takes on any issue, the assumption that a candidate's positions on issues is determinative is a reasonable one. Indeed, in presidential primaries most candidates tend to be identified as "liberal," "moderate," or "conservative," based on their positions on a range of domestic and foreign policy questions.[22] In Chapter 4, however, I shall show that if there are multiple issues on which candidates are simultaneously evaluated, the simple one-dimensional spatial analysis heretofore described may not yield optimal positions that are in equilibrium.

NOTES

1. E. S. Stavely, *Greek and Roman Voting and Elections* (Ithaca, N.Y.: Cornell University Press, 1972).
2. A good treatment of spatial voting games, as well as an analysis of other complications that may be made in the simple unidimensional models developed here, is given in James M. Enelow and Melvin J. Hinich, *The Spatial Theory of Voting: An Introduction* (Cambridge: Cambridge University Press, 1984).
3. See V. O. Key, Jr., with the assistance of Milton C. Cummings, Jr., *The Responsible Electorate: Rationality in Presidential Voting, 1936-1960* (Cambridge, Mass.: Harvard University Press, 1966). The debate over the significance of issues versus other factors in campaigns, such as candidate evaluations and party images, that affect a voter's decision is aired in *Controversies in Voting Behavior*, 2d ed., ed. Richard G. Niemi and Herbert F. Weisberg

(Washington, D.C.: CQ Press, 1984). These different influences are sometimes hard to distinguish conceptually, much less measure. For example, it seems plausible to assume that the *origins* of candidate evaluations and party images arise, at least in part, from the issue positions of candidates and parties—perhaps as the voter remembers them from earlier elections.

4. In some states voters registered as independents, who are unaffiliated with any party but are still allowed to vote in one party's primary, often choose the primary of the party having the most hotly contested election.
5. An issue on which attitudes can be indexed by some quantitative variable, like "degree of government intervention in the economy," obviously better satisfies this assumption than an issue that poses an either-or question—for example, whether or not to develop a major new weapons system.
6. The spatial representation of voter attitudes and candidate positions developed here was first used in Anthony Downs, *An Economic Theory of Democracy* (New York: Harper and Row, 1957).
7. Alternative models posit candidates who have policy preferences and view winning as a means to implement them, rather than candidates who more cynically adopt policy positions as a means to winning. A review and synthesis that combines both approaches is given in Donald Wittman, "Candidate Motivation: A Synthesis of Alternative Theories," *American Political Science Review* 77 (March 1983): 142-157.
8. I assume for now that A does not suffer any electoral penalty at the polls from changing his or her position, though fluctuations along the continuum may evoke a charge of being "wishy-washy," which is a feature of candidate positions analyzed in Section 3.7. Alternatively, the "movements" discussed here may be thought to occur mostly in the minds of the candidates before they announce their actual positions.
9. Steven J. Brams and Philip D. Straffin, Jr., "The Entry Problem in a Political Race," in *Political Equilibrium,* ed. Peter C. Ordeshook and Kenneth A. Shepsle (Boston: Kluwer Nijhoff, 1982), pp. 181-195. The median remains an equilibrium position when there are $n > 2$ candidates under "approval voting," which allows voters to vote for as many candidates as they like (without ranking). This is easy to show. Assume that any single candidate unilaterally departs from the median. Then a majority prefer the $n - 1$ candidates remaining at the median and will vote for all of them, presumably electing each with probability $1/(n - 1)$. This is tantamount to the two-candidate median-voter equilibrium result under plurality voting, except that one of several median candidates is elected under approval voting, whereas the single median candidate is elected under plurality voting (when the other departs from the median). See Gary W. Cox, "Electoral Equilibria under Approval Voting," *American Journal of Political Science* 29 (February 1985): 112-118. A general analysis of approval voting is given in Steven J. Brams and Peter C. Fishburn, *Approval Voting* (Boston: Birkhäuser, 1983).
10. For proofs, see Brams and Straffin, "The Entry Problem in a Political Race."
11. Palfrey argues that the stability of the two-party system in the United States may be explained by the fact that the two major parties, anticipating the

possible entry of a third party, deliberately position themselves away from the median to discourage entry. He shows that if voters are uniformly distributed, when the two parties take positions at 1/4 and 3/4—whereby 1/4 of the voters are to the left of the 1/4 party, and 1/4 to the right of the 3/4 party—a third party can do no better than win 25 percent of the vote, which is the maximum barrier that the two parties can erect against third-party entry. Thomas R. Palfrey, "Spatial Equilibrium with Entry," *Review of Economic Studies* 51 (January 1984): 139-156.

12. For a rational-choice analysis of this question, see Steven J. Brams, *Paradoxes in Politics: An Introduction to the Nonobvious in Political Science* (New York: Free Press, 1976), pp. 126-135, and references cited therein.
13. Hugh A. Bone and Austin Ranney, *Politics and Voters* (New York: McGraw-Hill, 1976), p. 81.
14. Ibid.
15. If the moderate candidate's position were at the median, he or she would receive more than half the votes between the points L and C, since voters are more concentrated around the median than at L or C. But if the moderate is to the right of the median, the votes that would be divided between him or her and the liberal candidate at the point midway between L and M would, if the moderate were sufficiently far from Md/Mn, give the advantage to the liberal candidate. The conservative candidate would get fewer votes the closer the moderate candidate approached, but, depending on the distribution, it is certainly possible that the liberal and conservative could both beat the moderate in the three-way contest depicted in Figure 3.8.
16. In some states, these entry choices are made not by the candidates but by a state official who places the names of all recognized candidates on the ballot, whether they have formally announced their candidacies or not. In other states, there are filing dates that must be met to appear on the ballot. But even these can be ignored in most states if a candidate runs as a write-in. However, successful write-in campaigns, especially by nonincumbents, are rare.
17. Solutions to examples involving both discrete and continuous distributions are given in Steven J. Brams, *Spatial Models of Election Competition*, Undergraduate Mathematics and Its Applications Project Expository Monograph Series (Newton, Mass.: Education Development Center, 1979; reprint, Lexington, Mass.: COMAP, 1983); and Brams and Straffin, "The Entry Problem in a Political Race."
18. See Brams, *Paradoxes in Politics,* pp. 53-65, and references cited therein.
19. The classic study is Murray Levin, *The Alienated Voter: Politics in Boston* (New York: Holt, Rinehart and Winston, 1960). For more recent analyses, see James D. Wright, *The Dissent of the Governed: Alienation and Democracy in America* (New York: Academic Press, 1976); and the several articles under "Political Alienation in America," in *Society* 13 (July/August 1976): 18-57.
20. Three out of five supporters of Eugene McCarthy in the Democratic party primary in New Hampshire in 1968 believed that the Johnson administration was wrong on Vietnam because it was too dovish rather than too hawkish;

these voters' attitudes represented a complete inversion of McCarthy's antiwar views. Richard M. Scammon and Ben J. Wattenberg, *The Real Majority: An Extraordinary Examination of the American Electorate* (New York: Coward, McCann and Geoghegan, 1970), p. 91.

21. Riker and Ordeshook draw a similar distinction between "indifference" and "alienation," though they use the former concept to refer to a "cross-pressured" voter, not to one who simply lacks information. See William H. Riker and Peter C. Ordeshook, *Introduction to Positive Political Theory* (Englewood Cliffs, N.J.: Prentice-Hall, 1973), pp. 323-330.
22. For empirical evidence that a single liberal-conservative dimension is no longer adequate to characterize the behavior of American voters and that two dimensions are needed (tapping questions of both economic policy and civil liberties), see William S. Maddox and Stuart A. Little, *Beyond Liberal and Conservative: Reassessing the Political Spectrum* (Washington, D.C.: Cato Institute, 1984).

4 Voting Paradoxes and Problems of Representation

4.1 Introduction

So far I have focused on strategic calculations, from the Bible to presidential elections, to argue that political actors are by and large rational. In this chapter, I shall shift the emphasis somewhat from the rationality of individuals', or groups of individuals', making calculations to what may be considered the rationality of institutions that channel conflict.

Of course, institutions as such do not make calculations. Yet institutional designers, in setting up democratic procedures, presumably seek to establish means for resolving conflicts that give equitable outcomes and have other desirable properties as well. To the extent that they are successful in achieving their goals, it is appropriate to characterize their efforts—and perhaps the institutions themselves—as rational. Recently, however, it has been demonstrated that the problems of designing robust democratic institutions, such as voting procedures that are relatively invulnerable to strategic manipulations, are indeed formidable.

4.2 The Paradox of Voting

The paradox of voting has probably been the source of more intellectual controversy and confusion than any other paradox in the social sciences, at least as judged by the amount that has been written about it. First discovered in the eighteenth century, it was largely forgotten until the pioneering work of Black and Arrow in the late 1940s and early 1950s resurrected interest in it and its ramifications. Since then, an enormous, though largely abstruse and mathematical, literature has developed concerning this paradox.

The consuming interest in this paradox may be attributed to several factors. First, in its most elementary form it can be simply stated, as can some of the most famous unsolved problems in mathematics. Second, this paradox is not only puzzling but also pathological—it says that there are fundamental problems connected with the idea of social or collective choice. Third, the implications of the paradox are seen by many analysts (though by no means all) as pervasive, potentially affecting all political choices made by democratic means.

The bulk of this chapter is devoted to a theoretical analysis of the paradox of voting and a consideration of its empirical manifestations; in addition, specific types of voting paradoxes that can arise under preferential voting systems that allow voters to rank candidates from best to worst will be briefly considered. Finally, paradoxes related to achieving fair representation—of minorities under a system of proportional representation, and of states (or other units) under a system of apportionment—will be illustrated.

Of course, what one regards as paradoxical is somewhat in the eye of the beholder. Not everyone will find each of the paradoxes discussed in this and subsequent chapters equally surprising or nonintuitive, although certainly some people have considered what I have labeled paradoxes puzzling if not deeply disturbing. To ascertain whether the questions raised by these paradoxes are of general interest and whether their analysis casts new light on enduring issues, a number of different models will be developed.

The standard example of the paradox of voting involves three voters who rank a set of three alternatives $\{x,y,z\}$ as shown in Table 4.1. Assume that all voters have *transitive* individual preference scales: a voter who prefers x to y and y to z will necessarily prefer x to z. The fact that this voter prefers x to y to z is indicated by preference scale (x,y,z).

Table 4.1 Paradox of Voting

Voter	Preference Scale
1	(x,y,z)
2	(y,z,x)
3	(z,x,y)

The paradox arises from the fact that if all voters have transitive preference scales, the social ordering nevertheless is intransitive: although a majority (voters 1 and 3) prefer x to y and a majority (voters 1 and 2) prefer y to z, a majority (voters 2 and 3) prefer z to x. This means that given at least three alternatives, there may be no social choice that is a *Condorcet candidate*—that is, one that can defeat every other alternative in a series of pairwise contests.

In the present example, every alternative that receives majority support in one contest can be defeated by another alternative in another contest. For this reason, the majorities that prefer each alternative over some other in a series of pairwise contests are referred to as *cyclical majorities:* they are in a cycle, $x > y > z > x$ (where "$>$" indicates "defeats"), that returns to its starting point. The absence of a clear-cut winner, or social choice, suggested by the paradox is not dependent, however, on a specific decision rule like majority rule (here, two votes out of three), as will be shown in subsequent sections.

Cyclical majorities may manifest themselves in various forms, including voting on candidate platforms. Consider a situation in which the voters do not vote on the basis of candidate positions on single issues, as assumed in the earlier one-dimensional spatial analysis (Chapter 3), but instead must choose among candidates who take positions on two or more issues.[1] For example, one issue might concern how a candidate would deal with inflation, another with superpower conflict.

To analyze some elementary consequences of multi-issue campaigns, consider a simple example of a campaign in which there are just two issues, X and Y. Assume that each candidate can take only one of two positions on each issue (that is, for or against), which are designated as x and x′, y and y′. Altogether, there are four possible *platforms,* or sets of positions on both issues, that a candidate can adopt: xy, x′y, xy′, or x′y′.

Assume that the electorate consists of three voters, and their preferences for each of the platforms are as shown in Table 4.2.[2] For each voter, the first platform in parentheses is most preferred, the second next-most preferred, and so on.

Table 4.2 Preferences of Three Voters for Platforms

Voter	Preference Scale
1	(xy,xy′,x′y,x′y′)
2	(xy′,x′y′,xy,x′y)
3	(x′y,x′y′,xy,xy′)

Assume that there are just two candidates, and one is elected if a majority of voters (two out of three) prefer his or her platform to that of the other candidate. What platform should a candidate adopt if his or her goal is to get elected?

To answer this question, one might start by determining which position on each issue would be preferred by a majority if the issues were voted on separately. Since x is preferred to x′ by voters 1 and 2 and y is preferred to y′ by voters 1 and 3 (compare the first preferences of voters in Table 4.2), it would appear that platform xy represents the strongest set of positions for a candidate.

But this conclusion is erroneous. Despite the fact that a majority would prefer positions x and y if the issues were voted on separately, only one voter prefers the *combination* xy to x′y′; since the latter platform is preferred by a majority (voters 2 and 3), it defeats platform xy. Thus, a platform whose individual positions (x and y) are favored by majorities may be defeated by a platform containing individual positions (x′and y′) that only minorities favor. In recognition of this possibility, Downs argued that it may be rational for candidates to construct platforms that appeal to "coalitions of minorities." [3]

In this example a paradox of voting underlies the divergence of less-preferred individual positions and a more-preferred platform that combines them.[4] Here there is no platform that can defeat all others in a series of pairwise contests. As shown by the arrows in Figure 4.1, which

Figure 4.1 Cyclical Majorities for Platform Voting

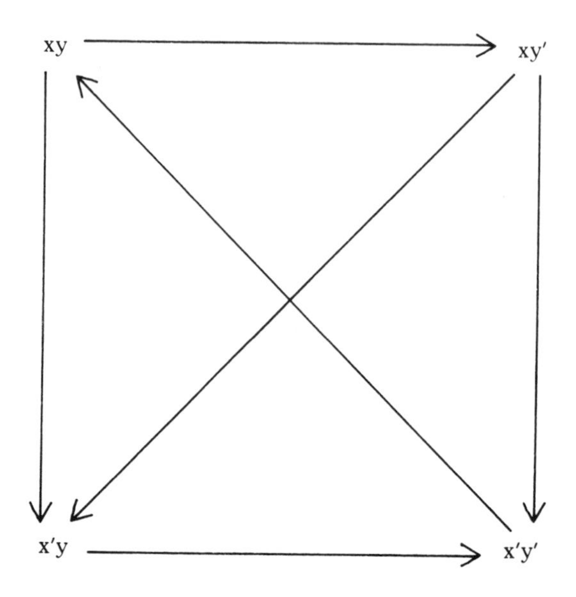

indicate majority preferences between pairs of platforms, cyclical majorities crop up again: every platform that receives majority support in one contest can be defeated by another majority in another contest.

The main conclusion derived from this simple platform example is that there may be no invulnerable set of positions a candidate can adopt on two (or more) issues: any set of positions one candidate takes can be defeated by a different set adopted by another candidate. Therefore, without any shift in the preferences of voters, a candidate running on a given platform could win an election in one year and lose it in the next, depending on an opponent's positions.[5]

No wonder candidates make such efforts to anticipate an opponent's positions so that they can respond with a set that is more appealing to the voters. Of course, some candidates try to avoid this problem by being intentionally vague about their positions in the first place, as Downs pointed out.[6] But this strategy of ambiguity may lead to its own problems, as was shown in Section 3.7.

By now it should be evident why elections so often seem to yield capricious outcomes. The strongest theoretical result of Chapter 3—that the median position is stable and optimal in a two-candidate election—can be undermined if there is more than one significant issue on which candidates take positions. Indeed, no set of positions will be stable if there exists a paradox of voting, nor will any set be optimal in the sense of guaranteeing a particular outcome whatever the positions of an opponent. In fact, contrary to expectations, a candidate's best set of positions on issues in a race may be the *minority* positions on the issues considered separately, depending on the positions of an opponent.

These findings do not depend on the exact nature of the underlying distributions of attitudes of voters or the precise location of candidates with respect to these distributions. They depend only on qualitative distinctions—dichotomous positions of candidates (either pro or con on an issue), ordinal preferences of voters (rankings of alternatives)—and are, therefore, of rather general theoretical significance whatever the quantitative characteristics of a race are.

Probably the best advice to take from the analysis in this section is negative: avoid reading too much into spatial analysis based on a single issue if there may be other issues of importance in a campaign. Multiple issues greatly complicate—and may ultimately confound—single-issue spatial analysis, which always shows that voters prefer one position to every other in pairwise contests (except for ties); hence, there can be no paradox of voting. Despite such complications, it is important to try to link candidate positions and voter attitudes, and spatial analysis provides a useful framework within which to do so for both single and sequential elections (including primaries).

4.3 Arrow's General Possibility Theorem

Consider what the paradox of voting says. Clearly it demonstrates, at least as manifested in the previous examples, that a concept of rationality based on transitivity cannot be transferred directly from individuals to a collectivity via some decision rule like that of simple majority. Something queer happens on the way, something comes apart, which complicates the idea of a social choice based on this kind of internal consistency. Riker and Ordeshook have commented on the qualitative difference between individual and social decisions:

> Social decisions are not the same kind of thing as individual decisions, even though the former are constructed from the latter. As a consequence of that difference, social decisions are sometimes arbitrary in a way personal decisions are not; personal decisions follow from persons' tastes, but social decisions do not follow from the taste of society simply because it is never clear what the taste of society is.[7]

In other words, there seems to be something awry with the very idea of a coherent social preference or social choice. When is it possible, then, for society to construct a transitive social preference order from the preference scales of its individual members?

Arrow's approach to this question was not to analyze a specific method for summing individual preferences but rather to postulate a set of abstract conditions that *any* summation method should meet.[8] He then asked what implications summation methods satisfying these conditions have for the *collective rationality* of social outcomes—specifically, whether they lead to transitive social choices—given that the preferences of individuals are transitive.

To ensure the transitivity of individual choices, Arrow made two assumptions about the relation by which individuals order pairs of alternatives, indicating either preference or indifference. Letting R symbolize this relation (read "is preferred to" or "is indifferent to"), the assumptions are:

(a) *Connectivity.* For any two alternatives x and y, either xRy or yRx. That is, the alternatives (for example, political candidates) possess some common property (for example, positions on an issue) that allows individuals to compare them with respect to their values.

(b) *Transitivity.* For any three alternatives x, y, and z, if xRy and yRz, then xRz. That is, individuals are consistent (or transitive) in their ranking of alternatives.

Given that there exist connected alternatives that individuals can transitively order, Arrow postulated five conditions that he believed would render social choices democratic. These conditions may be thought of as reasonable requirements that any method of aggregating or summing individual preferences into a social outcome should satisfy, where the social outcome is simply society's ranking of the alternatives. Very roughly, Arrow's conditions are:

1. *Universal admissibility of individual preference scales.* All possible orderings of alternatives by individuals are admissible; no institutions (for example, political parties) can restrict the orderings so that certain preference scales cannot be expressed.
2. *Positive association of individual and social values.* Given that xRy is the social ordering, if individuals either raise or do not change the ranking of x in their preference scales and the ranking of y remains unchanged, it is still the case that xRy. This restriction ensures that the method of summing individuals' preference scales reflects, in a nonperverse way, these preferences: the social ranking of x does not respond negatively to changes in rankings by individuals.
3. *Independence from irrelevant alternatives.* If S is a subset of the set of available alternatives and the preference scales of individuals change with respect to alternatives not in S, then the social ordering for alternatives in S does not change. (This has been by far the most controversial of Arrow's conditions; I shall discuss its implications in detail in Section 4.4.)
4. *Citizens' sovereignty.* For any two alternatives x and y, there exist individual preference scales such that x is preferred to y in the social ordering. In other words, the social outcome is not imposed; at the extreme, if all individuals should prefer x to y, x cannot be prohibited as the social outcome. This condition, in effect, outlaws the possibility that a social outcome is unrelated to the preference scales of society's members.
5. *Nondictatorship.* For any two alternatives x and y, there is no individual such that whenever he or she prefers x to y, x is always preferred to y in the social ordering. In other words, there is no individual who can dictate the social ordering of alternatives. Thus, by this condition an alternative cannot be prescribed from the *inside* (for instance, by a dictator), just as by condition 4 an alternative cannot be imposed from the *outside* (for example, according to some societal ethic).[9]

The intent of these conditions is to link society's ordering of alternatives to individuals' preference scales in a nonarbitrary way—that is, in a way that would make the social outcome responsive to the preference scales of individuals (as spelled out in conditions 2-5), whatever they may be (as guaranteed by condition 1). Remarkably, there is *no* method of summing individual preferences that satisfies all these conditions and yields a *social ordering* satisfying assumptions (a) and (b), previously postulated for individuals. Every method of summation that satisfies the five conditions does not produce a connected and transitive social ordering; if conditions 1-3, connectedness, and transitivity are satisfied, either condition 4 (citizens' sovereignty) or condition 5 (nondictatorship) must be violated. If cyclical majorities are to be avoided, therefore, the social outcome must be either imposed or dictated.

This is the essence of Arrow's general possibility—sometimes called impossibility—theorem. Although the proof of this famous theorem is beyond the scope of this book, there is no difficulty in explaining its import. In effect, it condemns to an ineradicable arbitrariness *all* methods of summing individual preferences that satisfy the notions of fairness and justice embodied in the five conditions. At least for some preference scales of individuals (by condition 1, none can be precluded), any method of summation that meets the other four conditions will lead necessarily to cyclical majorities, thus making the social outcome dependent on which pair of alternatives is compared. Hence, the paradox of voting cannot be dismissed as an aberration of majority rule or any other method of summing individual preferences.

The strength of Arrow's abstract formulation of the five conditions is that he did not have to examine every summation method that has ever been proposed or is ever likely to be proposed. By abstractly defining a *class* of methods, rather than concretely defining a particular summation procedure, he was able to administer the *coup de grâce* to all procedures that possess the properties of this class. Given the defining properties of this class (Arrow's five conditions), an arbitrary social choice cannot be avoided, which renders the notion of social consensus achieved by procedures having these properties virtually meaningless. Concretely, the "consensus" cannot be a Condorcet candidate, for such procedures of necessity lead to social preferences that cycle.

4.4 Relaxing Arrow's Conditions

Perplexed and dismayed by the negativeness of Arrow's result, many theorists, not surprisingly, have attacked the reasonableness of certain of his conditions. This attack has led to various refinements in Arrow's statement of the original conditions, new axiomatizations of social choice, and many new theorems.[10] But except for one minor flaw found in Arrow's original proof (discussed by Arrow in the second edition of *Social Choice and Individual Values*), his basic impossibility result has remained intact for a generation.

If the correctness of Arrow's theorem has proved logically unassailable, there nevertheless remains a good deal of controversy over its interpretation and its significance for the study of social and political processes. Since the conditions in Arrow's proof of the general possibility theorem are all necessary, and together sufficient, to ensure the occurrence of the voting paradox, one way to avoid the paradox is to drop one of the conditions. The most controversial, and some have argued the most unrealistic, of the conditions is condition 3, independence from irrelevant alternatives.

It is not hard to construct an example in which the independence condition is violated. If the preference scales of five voters change from (1) (x,y,z) to (x,z,y), (2) (x,y,z) to (x,z,y), (3) (y,z,x) to (z,y,x), (4) (y,z,x) to (y,x,z), and (5) (z,x,y) to (x,z,y), no change occurs in each voter's ranking of alternative x vis-à-vis alternative y. Yet under a scoring system called the *Borda count*—whereby weights of 3, 2, and 1 are assigned to first, second, and third choices, respectively, and that alternative with the highest score wins—the first set of preference scales yields 10 votes for x versus 11 votes for y, whereas the second set of preference scales yields 12 votes for x versus 8 votes for y.

Thus, the voters' ranking of alternative z is *not* irrelevant to the social choice—y in the case of the first set of preferences, x in the case of the second—for the subset S containing alternatives x and y. That is, if S = $\{x,y\}$ in condition 3, the Borda count method violates condition 3 because the social ordering of x and y does change, even though the ordering of x and y in *each voter's* preference scale does not: each voter continues to prefer x to y, or vice versa, but the relationship of one or both of these alternatives to z changes from the first to the second rankings.[11]

Should one be greatly upset by this lack of independence? Or is there little need to be alarmed that the social choice between two alternatives may be sensitive to voters' rankings of a third? Before discussing this issue, I shall show how the relaxation of another of Arrow's conditions may provide a different possible escape from the paradox of voting, thereby enabling one to avoid the consequences of social incoherence endemic to any summation method that fulfills all the conditions of the general possibility theorem.

Even before Arrow proved his famous theorem, Black showed in a series of articles in the late 1940s that by imposing certain restrictions on individuals' preference scales (which, of course, violates condition 1, universal admissibility of individual preference scales), an outcome favored by a majority can be ensured.[12] Given a decision rule of simple majority, one obvious condition sufficient to guarantee that a majority of voters favor a particular outcome is that more than half of all individuals rank that outcome highest on their preference scales. However, this condition is highly restrictive and, except when individual preferences are very homogeneous, is probably rarely satisfied in practice.

In fact, to ensure that one outcome is preferred by a majority to all others, one need insist only that the preference scales of individuals be "single peaked." Black defined *single peakedness* in terms of a coordinate system in which the outcomes or alternatives are arranged along the horizontal axis and the preference rankings by individuals of these alternatives along the vertical axis. The preference scale of each individual can then be represented geometrically by a set of connected lines, or curves, as illustrated in Figure 4.2.

In this example, individual 1 ranks alternative x first, alternative y second, and alternative z third, so his or her preference scale is (x,y,z). Similarly, the preference scales of individuals 2 and 3 are (y,x,z) and (z,y,x), respectively. Note that each individual's preference curve in Figure 4.2 (1) has a single peak and (2) always slopes downward from this peak in one or both directions.

In Figure 4.3, by contrast, not all individual preference scales are single peaked; individual 3's preference curve first slopes downward from

Figure 4.2 Geometric Representation of Single-Peaked Preference Scales of Three Individuals

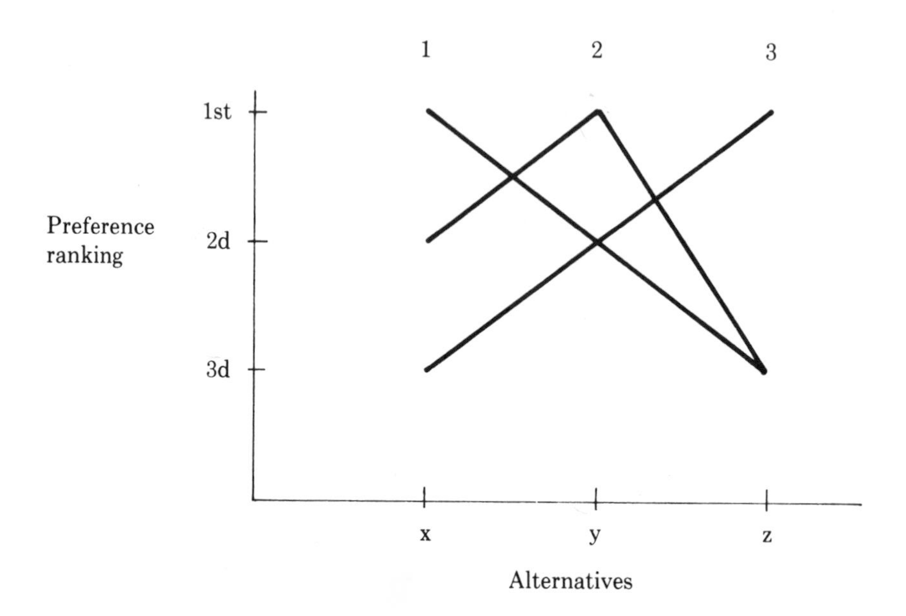

Figure 4.3 Nonsingle-Peaked Preference Scales

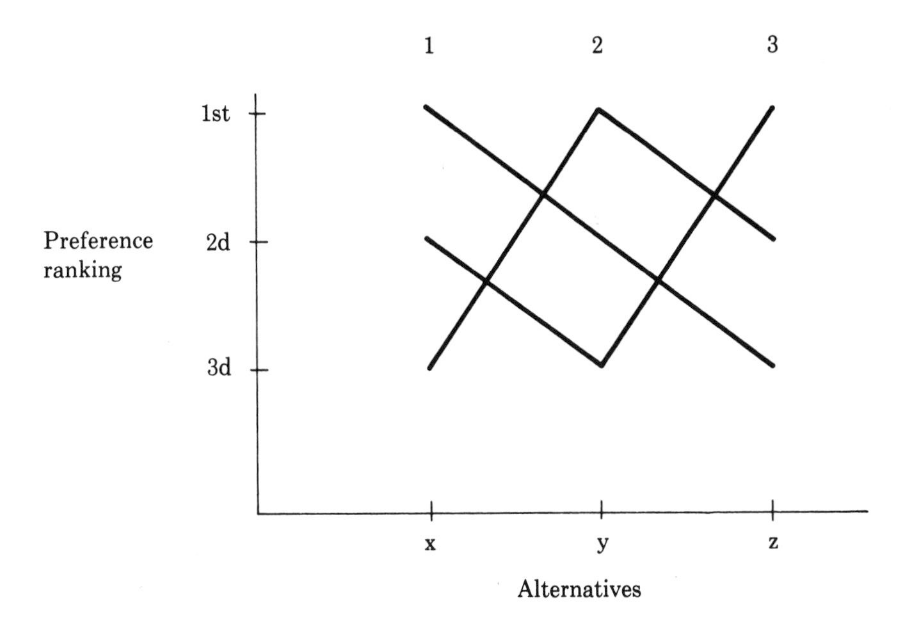

his or her most-preferred alternative (z) and then changes direction and slopes upward (reading from right to left). Only when an individual's preference curve increases up to a peak—or is at a peak immediately—and then decreases after the peak is his or her preference scale said to be single peaked.

The preference scales of a *set* of individuals are said to be *single peaked* if there is *some* arrangement of alternatives along the horizontal axis such that the preference curves of *all* individuals are single peaked. The reader can check that no matter how the alternatives in Figure 4.3 are ordered along the horizontal axis (there are six possible arrangements), there is *no* ordering that renders the preference curves of all individuals single peaked.

Black demonstrated that if the preference scales of all individuals can be represented by a set of single-peaked preference curves, then there is one alternative that is preferred by a majority to all others (that is, is a Condorcet candidate). Thus, single peakedness ensures that there are no cyclical majorities and the social ordering is transitive. In the case of the preference scales represented in Figure 4.2, for example, alternative y is preferred to x by individuals 2 and 3 and to z by individuals 1 and 2. Thus, alternative y is the social choice; it is also easy to show that a majority (individuals 1 and 2) prefer alternative x to alternative z, so the social preference order is (y,x,z).

In general, for a set of single-peaked preference curves, the social choice will be the alternative at the peak of the *median curve,* or the curve to the left of whose peak lie exactly as many preference curve peaks as to the right. Obviously, if the number of persons is even, there is no single median curve and the possibility of ties exists. In this case, therefore, there may be no single alternative preferred by a majority to all others, though there will be one that is undefeatable if all preference curves are single peaked.

Single peakedness provides a useful criterion for determining whether the preference scales of individuals are sufficiently similar, or homogeneous, for there to be an alternative that defeats or—in the even case—is undefeatable by all other alternatives in a series of pairwise contests. Moreover, if there is an odd number of individuals, one can tell which alternative—that coincident with the median preference curve—is the Condorcet candidate.

Roughly, single peakedness can be interpreted to mean that there exists a single dimension underlying the preferences of all individuals (for example, a liberalism-conservatism scale) along which the alternatives (for example, political candidates) can be ordered. The existence of such a dimension *is* consistent with individuals' having diametrically opposed preference scales; it implies only that they exercise a uniform standard in ordering the alternatives.

Methods for determining the "degree" of single peakedness of a set of preference scales have been developed and applied to empirical data. In Section 4.7 some empirical instances of the paradox of voting wherein the single-peakedness condition was apparently not met will be discussed. At this point, however, it is helpful to consider an alternative characterization of single peakedness that sheds considerable light on the nature of individual preference scales that give rise to the voting paradox. This characterization will also facilitate the analysis of the probability of occurrence of the paradox in Section 4.5.

From the preceding discussion and examples, it is apparent that there may be a Condorcet candidate even when the preference scales of individuals differ markedly. For example, for the preference scales depicted in Figure 4.2, every person most prefers a different alternative (that is, the peak of each preference is located at a different point on the horizontal axis). Nevertheless, a majority still prefer alternative y to either other alternative. This raises a question: How heterogeneous must individual preference scales be for the paradox of voting to occur?

Looking at the preference scales of individuals in Figure 4.3, where there is a paradox, one finds that not only do the first preferences of each individual differ but so do the second preferences: the second choice of individual 1 is alternative y; of individual 2, alternative z; and of individual 3, alternative x. Given that the first and the second choices of all three individuals differ, it necessarily follows that the least-preferred alternatives of the individuals also differ, which a glance at Figure 4.3 confirms. Situations wherein each alternative is ranked differently by every individual—first on one person's preference scale, second on another's, and last on the third's—always result in the paradox.

To characterize situations generally in which the paradox cannot occur, Sen defined a set of preference scales to be *value restricted* if all individuals agree that there is some alternative that is never best, medium, or worst for every set of three alternatives, or "triple." Thus, the triple in Figure 4.2 is value restricted since alternative y is not worst on anyone's preference scale. Given that a set of alternatives is value restricted, Sen proved that majority rule satisfies all Arrow's conditions [and assumptions (a) and (b)], except, of course, condition 1, universal admissibility of individual preference scales.[13]

It is evident, then, that the paradox of voting can be circumvented if individuals' preference scales reflect a certain modicum of similarity. Since the preferences of citizens in democracies and other social collectivities to which conditions 2-5 of Arrow's theorem might apply undoubtedly exhibit some similarities, an evaluation of the real-world implications of the paradox of voting must take account of its expected frequency of occurrence. If this is very small, the paradox can quickly be dismissed as a curious but unimportant phenomenon. If the paradox has a high probability of occurring, on the other hand, it may have serious ramifications for the social coherence of democratic societies.

4.5 Probability of the Paradox of Voting

Make the simplifying assumption that all orderings of alternatives are equally likely for every individual. Then it is not difficult to calculate the probability of the paradox for three individuals who choose among three alternatives. There are six ways in which each individual may order the set of three alternatives $\{x,y,z\}$:

1. (x,y,z)
2. (x,z,y)
3. (y,x,z)
4. (y,z,x)
5. (z,x,y)
6. (z,y,x)

Hence, there are $6^3 = 216$ ways in which three individuals can express their preferences.

For a social intransitivity to occur, as noted in the discussion of value restrictedness (Section 4.4), neither the first, second, nor third preferences of the three individuals can agree. Thus, once one individual has chosen a preference scale from among the six possibilities [for example, (x,y,z)] a second individual has a choice between the only two preference scales [(y,z,x) and (z,x,y)] on which none of the alternatives is ranked in the same position as on the first individual's preference scale. Given the choice of the first two individuals, the third individual is limited to just one preference scale.

Consequently, given three alternatives, there are $(6)(2)(1) = 12$ ways in which three individuals can choose preference scales that will yield an intransitive social preference. The probability, therefore, of a social intransitivity is $12/216 = 0.056$, assuming all preference scales are equiprobable.

The probabilities that no alternative will be preferred to all others for three or more individuals choosing among three or more alternatives are shown for several cases in Table 4.3. The probability values reveal an interesting pattern as the number of individuals and alternatives increase. For a fixed number of alternatives, the values increase—most rapidly in the beginning—as the size of a voting body approaches infinity, but not generally by a great amount. In the case of three alternatives, for example, the probability of the paradox increases from 0.056 to 0.080, or to less than 1 chance in 12. For a fixed number of individuals, on the other hand, the probability of the paradox always approaches 1.00 as the number of alternatives increases. Clearly, the occurrence of the paradox is much more sensitive to the number of alternatives than to the number of individuals.

Table 4.3 Probabilities of No Condorcet Candidate

Number of Alternatives	Number of Individuals						
	3	5	7	9	11	...	Limit
3	0.056	0.069	0.075	0.078	0.080	...	0.080
4	0.111	0.139	0.150	0.156	0.160	...	0.176
5	0.160	0.200	0.215	0.230	0.251	...	0.251
6	0.202	0.255	0.258	0.284	0.294	...	0.315
7	0.239	0.299	0.305	0.342	0.343	...	0.369
⋮	⋮	⋮	⋮	⋮	⋮	⋮	⋮
Limit	1.000	1.000	1.000	1.000	1.000	...	1.000

Source: Peter C. Fishburn, *The Theory of Social Choice*. Copyright © 1973 by Princeton University Press. Figure 8.2 reprinted with permission of Princeton University Press.

4.6 Empirical Examples of the Paradox of Voting

It is hard to evaluate the empirical significance of the paradox from the probabilistic calculations just outlined since they are based on an equiprobability assumption about individuals' preference scales, which in practice is probably violated. The question is: In what direction do deviations from this assumption occur?

Since political actors may have an incentive to contrive a paradox that exploits an apparent lack of consensus among voters, Riker and Ordeshook argue that the theoretical calculations probably understate the significance of the paradox.[14] Moreover, because the manipulation of outcomes requires a not inconsiderable amount of planning and effort, one would expect that the more important and controversial an issue is, the more likely political actors will be motivated to contrive a paradox that is not inherent in the distribution of preferences.

As evidence for contrivance, Riker gives two empirical examples.[15] The first concerns federal aid for school construction, which was considered by the U.S. House of Representatives in 1956 in terms of the following three alternatives: (1) an original bill, O, for grants-in-aid for school construction; (2) the bill with the so-called Powell amendment, A, which provided that no federal money be spent in states with segregated public schools; and (3) no bill, N, the status quo. As Riker reconstructed the situation, there were basically three groups of voters with the following preference scales:

1. The passers (mostly southern Democrats) favored school aid, but they were so repelled by the Powell amendment that they preferred no bill to the amended bill. Thus, their preference scale was (O,N,A).
2. The amenders (mostly northern Democrats) were prointegration, but if the Powell amendment were not successful, would still prefer school aid to no action. Thus, their preference scale was (A,O,N).
3. The defeaters (mostly Republicans) preferred the status quo over either version of the bill, with the prointegration bill apparently more palatable than the original bill. Thus, their preference scale was (N,A,O).

The voting procedure used in Congress (and in many other voting bodies) is a binary procedure called the *amendment procedure,* whereby the set of outcomes is divided into two subsets, and each subset in turn into two further subsets, as shown in the *voting tree* depicted in Figure 4.4 (read from top to bottom). In the example at hand, the procedure pits the subset of outcomes {A,N} against the subset {O,N}. If the amend-

Figure 4.4 Amendment Procedure

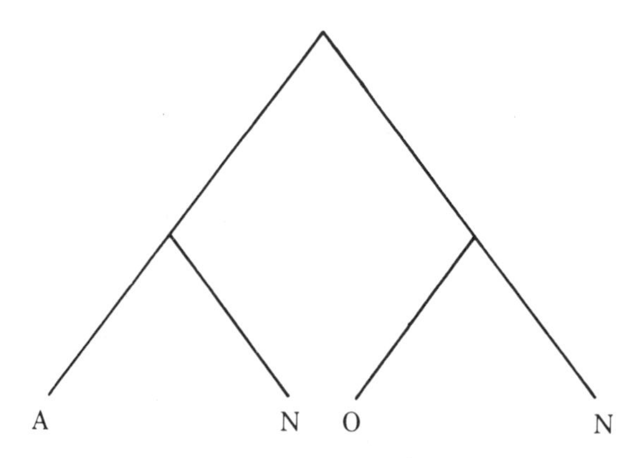

ment is adopted, the second vote is on the amended bill (A) versus no bill (N); if not, the second vote is on the original bill (O) versus N. Defeat can thus occur in two different ways under this procedure.

The first vote on the Powell amendment passed, but the second vote on the amended bill failed. The first roll call reveals that the northern Democrats needed (and got) the support of Republicans—whose first preference was defeat of any school-aid bill—on the Powell amendment. Riker's interpretation is that the Republicans, realizing that a majority probably favored the original bill, may have voted for the amendment {A,N} instead of {O,N} only so as to detach southern Democrat support on the vote for final passage (A versus N), causing it to fail. (On the other hand, if the Republicans had voted against the amendment initially, the vote for final passage would have been O versus N, and O would have won because both northern and southern Democrats would have supported it.)

Since the Republicans did not introduce the Powell amendment, however, it seems unfair to blame them for using it to defeat the school-aid bill. If there had been contrivance on their part, it would have arisen from not voting sincerely to affect a better outcome.[16] (*Sincere voting* is voting for more-preferred subsets over less-preferred subsets—in the case of the Republicans, for example, for {A,N} over {O,N} since they prefer A to O.) Yet in fact most Republicans voted sincerely for the amendment, consistent with their postulated preference scale.

In fact, it was the northern Democrats, and not the Republicans, who had the opportunity to contrive an outcome that would have been preferred by them over N, but they failed to exploit this opportunity. Specifically, if they had voted for {O,N} instead of their sincere choice

{A,N} on the amendment vote, {O,N} would have been selected. Then outcome O would have been the choice of the voting body on the final vote, since two of the three groups of voters (southern Democrats and northern Democrats) would have preferred it to outcome N.

Given the preference scales of the different groups of voters postulated in this example, therefore, it was not contrivance on the part of the Republicans that sank the federal school-aid bill but rather a failure to contrive (by voting *strategically*, or insincerely to effect a better outcome) on the part of the northern Democrats. Apparently, this possibility for contrivance was recognized by Rep. William L. Dawson (an Illinois Democrat), who was alone among black representatives in voting against the Powell amendment.

In 1957, a year after the Powell amendment had succeeded ({A,N} defeated {O,N}) but final passage had failed (N defeated A), the bill in its original form got majority support in the House (O defeated N). This history is convincing evidence that a paradox of voting actually existed: O defeats N, N defeats A, and A defeats O (or, equivalently, {A,N} defeats {O,N}; N is common to both subsets, which makes the effective comparison one between A and O, the noncommon members). But this analysis strongly suggests that the failure of either version of the school-aid bill to pass in 1956 was not a contrivance of the Republicans, at least in terms of the way in which they voted on the three alternatives. Furthermore, since they did not propose the amendment, it is hard to accuse them of splitting the Democratic opposition; the fault seems, rather, to lie with the Democrats themselves.

In his second example of the paradox, Riker presents a more convincing example of an amendment being specifically proposed to divide the support of the voters favoring the original bill. In this case, which occurred in the U.S. Senate in 1911, the original bill (O) was a proposal to amend the Constitution to provide for direct popular election of senators (which in 1913 became the Seventeenth Amendment). The amendment (A) provided that the guarantee in the original bill against federal regulation of these elections be deleted; by jeopardizing the ability of southern states to exclude blacks from the election process, it was specifically designed to antagonize senators from the South. The final alternative was the status quo, or no bill (N).

The preference scales of the southern Democrats, northern Democrats, and Republicans for {O,A,N} duplicate those in the previous case, with the sincere outcome N once again triumphant due to the failure of the northern Democrats to vote strategically for {O,N} initially and then for O. If there was contrivance on the part of the Republicans in introducing an amendment to drive a wedge through Democratic support, there was certainly no contrivance on the part of the northern Democrats to thwart this strategy. Although the idealist might characterize the voting behavior of the northern Democrats in these two cases as sincere, the realist might consider the epithet "naive" more fitting.[17] In cases like these, it is hard to believe that a knowledge of party positions, past voting, and other clues about likely voting behavior did not provide members with sufficient information on which to base predictions about outcomes—and, accordingly, to vote strategically.

Certainly in contemporary times, extensive debate and media coverage of controversial issues exposes the preferences of many members. Perversely, however, the availability of such information may inhibit strategic voting, for reasons that on reflection are not that surprising.

First, when preferences are more fully exposed, members will vote sincerely if the sanctions from their party, constituency, and other groups for taking the "wrong" (that is, insincere) public positions on roll-call votes override strategic voting considerations in Congress. Although the means used to achieve particular ends are ostensibly irrelevant to this analysis, in principle the costs associated with the diabolical aspects of strategic voting could be incorporated into the valuation of the ends themselves (for example, by distinguishing outcomes achieved by "devious" means from those achieved by "nondevious" means).

Another possible explanation for sincere voting might stem from the inability of members who do have something to gain from strategic voting to agree to coordinate their strategies. In situations wherein a party or coalition leader is skilled in parliamentary maneuver and persuasion, however, he or she may be able to convince a sufficiently large number of voters to take concerted action that has a disadvantageous immediate effect but will probably have a beneficial ultimate effect. Such a person seems to have been Lyndon B. Johnson, who as majority leader of the Senate was instrumental in the following case.[18]

In 1955 members of the Senate faced the following three alternatives: (1) an original bill, O, for an $18 billion highway program, which included the so-called Davis-Bacon clause that set up fair-pay standards for workers on federal construction projects; (2) the bill as amended, A, with the Davis-Bacon clause deleted; and (3) no bill, N, the status quo. Since the Davis-Bacon clause was anathema to southern Democrats, the voters, as before, divided into three groups.

1. Southern Democrats, with preference scale (A,N,O)
2. Northern Democrats, with preference scale (O,A,N)
3. Republicans, with preference scale (N,O,A)

Under the amendment procedure of Figure 4.4, sincere voting would result in N (as before) but strategic voting (by the northern Democrats) could induce A, which they preferred. The latter outcome eventually resulted because southern Democrats, led by Johnson, were able through parliamentary maneuvering to get the Davis-Bacon clause removed without a roll-call vote,[19] which was tantamount to getting northern Democrats to support them on {A,N} by voting strategically under the amendment procedure of Figure 4.4.[20] Then, on the vote pitting A against N, the northern Democrats joined the southern Democrats and supported A.

Farquharson quotes *Time* as reporting that the "Republicans flubbed their chance," presumably because their least-preferred outcome was adopted.[21] But if one assumes that the voters can communicate with each other for the purpose of coordinating joint strategies, the Republicans' problem is not easily solved. For whereas A, their worst outcome and the one adopted, is vulnerable to a coalition of northern Democrats and Republicans who support outcome O, O is vulnerable to a coalition of southern Democrats and Republicans who support outcome N; and N in turn is vulnerable to a coalition of northern and southern Democrats who support outcome A. In other words, once again there is a paradox of voting since there is no outcome that cannot be defeated—including that which the Republicans most favored, N. Thus, although the Republicans may have flubbed their chance, they had no surefire strategy for winning.

In the three instances of the paradox described in this section, the sincere outcome was chosen twice and the strategic outcome once.[22] To a certain extent, these outcomes are an artifact of the voting procedure used. If the voting procedure had been different, both the sincere and the strategic outcomes might have also been different.[23]

What these examples share, however, is the absence of a coalition that can form in support of an alternative and that will be resistant to coalitions supporting other alternatives. Given that groups of voters can communicate with each other, which obviously characterizes members of both houses of Congress, some group of members in any coalition that forms can be tempted to defect and join another coalition. This means that the outcome that is supported by a particular coalition and eventually prevails is very much an artifact of the stage in the voting process at which it forms. Assuming that it takes some time to mount a challenge to a coalition after it has formed, the later a coalition forms when there exists a paradox of voting, the more successful it will be in warding off such challenges.[24]

4.7 The Monotonicity Paradox

The arbitrariness that the paradox of voting introduces into the selection of alternatives—and even the very meaning of "social choice" when no alternative is preferred to every other—is only one of a set of problems that may arise in the elusive search for consensus and for procedures that will exhibit it (if it exists). The next example illustrates how information about voters' preference scales—or simply their first choices, as elicited, for example, by a poll—may create additional problems.[25] Assume that 17 voters divide into four classes, with the numbers that precede each class indicating the number of voters having each set of preferences over the set of alternatives {a,b,c}.

I. 6: (a,b,c)
II. 5: (c,a,b)
III. 4: (b,c,a)
IV. 2: (b,a,c)

If the voters choose their sincere strategies under plurality voting, an (accurate) poll would reveal alternatives a and b to be in a dead heat with 6 votes each (a is supported by class I voters, b by class III and class IV voters), with c trailing with 5 votes (class II voters). If this information elevates a and b to front-runner status and makes c appear to be out of the running, it would seem rational for the class II voters to vote insincerely for alternative a to prevent their worst choice, b, from winning, thereby ensuring the election of alternative a with 11 votes to 6 votes for b.

Now assume that all voters have complete information about each others' preference scales. Would this additional information cause any further adjustments in voter strategies in the resulting game?

Clearly, the class III voters, realizing that their sincere votes for b will help a, their least favorite, to win, could try to strike a deal with class II voters in support of c. If this deal were consummated, c would win with 9 of the 17 votes.

But then a coalition supporting b might form in opposition, and in turn a coalition supporting a might form in opposition to the b coalition. Suffice it to say that bargaining in this game would be complex, because there is a paradox of voting and hence no Condorcet candidate: a majority of 11 voters prefers a to b, a different majority of 10 voters prefers b to c, and a still different majority of 9 voters prefers c to a.

Would these cyclical majorities continue to create incoherence if the plurality election were followed by a runoff between the top two vote-getters? The answer is yes, because the contest would then revert to which pair gets into the runoff since the three possible pairs all lead to different winners.

Next suppose, for the sake of argument, that the class IV voters change their preferences and favor a over b rather than vice versa:

IV′. 2: (a,b,c)

In this case, a poll announcement, based on sincere voting by all four sets of voters under plurality voting, would indicate a (8 votes) and c (5 votes) to be the top two candidates. Given this information, the class III voters, favoring b, would be motivated to vote insincerely for c to prevent a from winning, so the effect of the announcement would be to give the edge to c by 9 votes to 8 votes.

In summary, alternative a wins after the poll announcement if the voters' preferences are as given at the beginning of this section, but if the class IV voters switch from IV to IV′—favoring a—c wins. This seems extraordinary: a wins when *fewer* voters favor him or her but loses when more voters favor him or her.

Obviously the poll has a strange effect, but the paradoxical result it produces may occur without the poll. If the plurality election were followed by a runoff, for example, and all voters voted sincerely, a would make the runoff and then *win* against b in it, given that the preference scale of the class IV voters was (b,a,c); but if the preference scale of the class IV voters were (a,b,c,), a would again make the runoff but this time *lose* to c in it.

What the runoff does, if the preferences of the class IV voters change in favor of a, is match a different pair. The poll announcement has the same effect as the runoff by making the contest appear to be one between different pairs, depending on the preferences of the class IV voters. Since which pair of candidates is matched determines the winner due to the paradox of voting (which persists if the preferences of the class IV voters change from IV to IV′), the election turns not on voter preferences alone but rather on how these preferences affect which two candidates are matched, either in a runoff or because of a poll announcement.

There is a *monotonicity paradox* if a winner (a in the previous example) is displaced after one or more voters (class IV voters in the example) change their preferences in a way favorable to him or her without changing the order in which they prefer other candidates. (A voting procedure is monotonic if this paradox can never arise.) Although new information (as from a poll) may give rise to a monotonicity paradox, this paradox is usually discussed in the context of voting systems that are vulnerable to it.

Plurality voting, followed by a runoff, is not the only system that is vulnerable to the monotonicity paradox and hence nonmonotonic. So is the Hare system of preferential voting, or "single transferable vote" as it is also known, which extends the idea of a runoff to allow for a sequence of elimination contests. Under this system, voters rank candidates; then, if no candidate receives a majority of first-place votes, the candidate with the fewest first-place votes is dropped and his or her second-place votes are transferred to the other candidates. (In a three-candidate election, this procedure is equivalent to conducting a runoff, under the assumption that supporters of the candidate who does not make the runoff will vote for their second choice in the runoff.)

If there are more than three candidates, candidates continue to be eliminated, with their second-place votes reallocated to the remaining candidates, until one candidate receives a majority of votes. The majority candidate is then declared the winner; still, as in a single-runoff election, it is possible for a candidate to win with, say, 10,000 first-place votes but to lose the election if, hypothetically, he or she should receive an additional 5,000 votes (from other voters who move him or her into first place without changing their ordering of other candidates) because of the monotonicity paradox. In a case like this, as Doron and Kronick imagine, many voters would probably be outraged by the election night report: "Mr. O'Grady did not obtain a seat in today's election, but if 5,000 of his supporters had voted for him in second place instead of first place, he would have won!" [26]

Unbelievable as this result may sound, the Hare system that can produce it is used today in public elections in such places as Australia; Ireland; Northern Ireland; Malta; Cambridge, Massachusetts; and New York City. Not only is the Hare system vulnerable to the monotonicity paradox, but it also does not ensure the election of a Condorcet candidate. In fairness to the Hare system, it should be noted that it has certain strengths in ensuring proportional representation (for example, of minorities) in multiple-winner elections, such as to a city council (Section 4.8); the criticism directed at the Hare system and runoff elections here is of their nonmonotonicity, especially in single-winner elections.

The existence of the monotonicity paradox had been suspected for a long time, but not until 1973 was it conclusively established.[27] Fishburn has shown that a number of ranking and sequential-elimination voting systems are vulnerable to the paradox, but most single-ballot systems—like plurality voting and approval voting, which do not involve the elimination of candidates and the transfer of votes—satisfy monotonicity.[28] As was shown earlier in the case of plurality voting, however, poll announcements may induce voters to make strategic calculations that engender the monotonicity paradox.

Under approval voting voters can vote for, or approve of, as many candidates as they like. Each candidate approved of receives one full vote, and the candidate with the most votes wins.[29] If all voters under this system indicated approval of only their first choices in a poll, the class II voters in the example at the beginning of this section would have an obvious incentive to switch for strategic reasons from strategy $\{c\}$ to $\{a,c\}$ as a result of the poll announcement favoring alternatives a and b (the candidates in braces are the subsets of candidates that voters in the class approve of). However, this would not affect the fact that alternative a would still go into the runoff against b and win.

Now consider what would happen if the preferences of the class IV voters changed from (b,a,c) to (a,b,c), favoring a as before. If, as before, all voters approved of only their top candidates, the poll would indicate a and c to be the top two candidates. The announcement of the poll would induce the class III voters to switch from {b} to {b,c}, but all this would do is solidify c's place in the runoff against a, and thereby c's victory; hence, a's ascent in the ranks of class IV voters succeeds only in inducing their worst alternative. Thus, poll announcements can render approval voting, as well as plurality voting, nonmonotonic insofar as they induce voters to vote *as if* there were a runoff.

Nonmonotonicity contradicts what would seem to be a basic principle of democratic elections: if some voters raise a candidate in their preference scales without changing their ordering of other candidates, and that candidate consequently receives more votes, he or she will *not* do worse in an election. Unfortunately, as has been shown, being elevated in the preference scales of some voters may sabotage a candidate's victory, depending on the system used for aggregating votes.

In my opinion, this pathology is most serious when it is the product of sincere voting, as it is under the runoff and Hare systems. By contrast, the ascent of a candidate in some voters' preference scales never hurts and sometimes helps under plurality and approval voting. The rub under these systems comes only when voters react strategically to certain information, such as given by a poll, that identifies the two front-runners.

In elections in which such information is generated, the monotonicity paradox is potentially a problem, though how serious is difficult to say. It should be recognized, however, that the information a poll provides may have pathological consequences through changing voter strategies (in such a way as to undermine favored candidates.[30]

4.8 Problems in Achieving Proportional Representation

Voting systems like Borda and Hare, by allowing voters to rank the candidates, enable minorities to ensure their approximate proportional representation (PR) on a council or in a legislature. Under Borda, a minority that ranks one candidate first can give him or her enough votes to be among the top several vote-getters and hence win a seat. Likewise, under Hare, if a candidate is ranked first by a minority of voters, he or she may be able to avoid elimination by obtaining at least the minimum of first-place votes needed to win a seat, which is called the quota.

The quota under the Hare system becomes a lower and lower threshold as the size of the voting body increases. Thus, for example, a 10 percent minority can generally win about 10 percent of the seats on a council or in a legislature if its supporters concentrate their votes on a subset of candidates commensurate with the minority's size in the electorate.[31]

Is it possible to ensure PR in multiple-winner elections in a way that avoids the nonmonotonicity of the Hare system? (Borda has other theoretical deficiencies—for example, it does not always lead to the election of Condorcet candidates—and is used in no public elections of which I am aware, whereas the types of systems discussed next are used and so seem better candidates for possible reform.) I shall describe a new nonranking voting system, which is called adjusted district voting (ADV) and is a variant of extant systems, and show that it has certain advantages over these systems but also may confound (to a degree) the achievement of PR.[32]

ADV nicely illustrates trade-offs that seem inevitable if one believes a representative democracy should provide some assurance of minority representation. A simplified form of this system can be modeled in terms of the following assumptions:

1. There is a jurisdiction divided into equal-size districts, each of which elects a single representative to a legislature.
2. There are two main factions in the jurisdiction, one majority and one minority, whose size can be determined. For example, if the factions are represented by political parties, their respective sizes can be determined by the votes that each party's candidates, summed across all districts, receive in the jurisdiction.
3. The legislature consists of all representatives who win in the districts *plus* the largest vote-getters among the losers, who are added to the legislature to achieve PR if it is not realized in the district elections. Typically, this adjustment involves adding minority-faction candidates who lose in the district races to the legislature so that the body mirrors the majority-minority breakdown in the electorate as closely as possible.
4. The size of the legislature is variable, with a lower bound equal to the number of districts (if no adjustment is necessary to achieve PR) and an upper bound equal to twice the number of districts (if a minority of nearly 50 percent wins no district seats).

ADV is in fact used in a limited form in Puerto Rico and has been proposed for use in Great Britain.[33] It is one of a class of what are variously called add-on, additional-member, and topping-up systems, probably the best known of which is West Germany's (variations can be found in Denmark, Iceland, and Sweden). In the German Bundestag, approximately half the seats are decided in single-member district contests, with the other half determined from second votes (for a national party) that are used, after the district results are factored in, to achieve PR (usually for minority parties).[34]

If there are more than two parties under an additional-member system, however, a fixed fraction of add-on seats might work well in one election but could prove quite inadequate in ensuring PR in the next election. How, for example, can a legislature that provides for a doubling in size of the original legislature to satisfy PR (like the Bundestag) accommodate a three-party national vote split of 40, 30, and 30 percent in which the largest party wins in every district? Patently, some of the 60 percent who voted for the two minority parties cannot be accorded PR if they together can be allowed no more than 50 percent of the seats.

An extension of ADV to multiparty systems offers a flexible approach to such a problem by making the size of the legislature variable within certain limits. But even in the two-party case, ADV runs afoul of a surprising problem.

Suppose there are eight districts in a jurisdiction. If there is an 80 percent majority and a 20 percent minority in the jurisdiction, the majority is likely to win all the districts unless there is an extreme concentration of the minority in one or two districts.

Suppose the minority wins no districts. Then its two biggest vote-getters could be given two extra seats to provide it with representation of 20 percent in a body of ten members, exactly its proportion in the electorate.

Suppose the minority wins one district, which would provide it with representation of about 13 percent in a body of eight members. If it were given an extra seat, its representation would rise to about 22 percent in a body of nine members, which would be closer to its proportion in the electorate. However, assume that the addition of extra seats can never make the minority's proportion in the legislature exceed its proportion in the electorate. Thus, with a victory in one district the minority party would be underrepresented at 13 percent.

Clearly, the minority would benefit by winning no seats and then being granted two extra seats to bring its proportion up to exactly 20 percent. To prevent a minority from benefiting by *losing* in district elections, assume the following contraint: the allocation of extra seats to the minority can never give it a greater proportion in the legislature than it would obtain had it won in more district elections.

How would this constraint work in the example? If the minority won no seats in the district elections, then the addition of two extra seats would give it 2/10 representation in the legislature, exactly its proportion in the electorate. But if the minority had won exactly one seat, it would *not* be entitled to an extra seat—and 2/9 representation in the legislature—because this proportion would exceed its 20 percent proportion in the electorate. Hence, the minority's representation would have to remain at 1/8—if it won in exactly one district—to satisfy the constraint.

Because $2/10 > 1/8$, the constraint prevents the minority from gaining two extra seats if it wins no district seats initially. Instead, it would be entitled in this case to only one extra seat—giving the minority one of nine seats—which satisfies the constraint because $1/9 < 1/8$. But 1/9, or about 11 percent, is only about half the minority's proportion in the electorate. In fact, Fishburn and I prove in the general case that the constraint may prevent a minority from receiving up to about half of the extra seats it would be entitled to on the basis of its proportion in the electorate.[35]

The constraint may be interpreted as a kind of "strategyproofness" feature of ADV: it makes it unprofitable for a minority party deliberately to lose in a district election to do better after the adjustment that gives it extra seats. But strategyproofness, in precluding any possible advantage that might accrue to the minority from throwing a district election, has a price. As the example demonstrates, it may severely restrict the ability of ADV to satisfy PR.

Thus, under ADV there is a conflict: one cannot guarantee a close correspondence between a party's proportion in the electorate and its representation in the legislature if one insists on the strategyproofness constraint. Dropping it allows one to approximate PR but may give an incentive to the minority party to lose in certain district contests in order to do better after the adjustment.

It is worth pointing out that the second chance for minority candidates under ADV would encourate them to run in the first place, because even if most or all of them were defeated in the district races, their biggest vote-getters would still get a chance at the (possible) extra seats in the second stage. But these extra seats might be cut by as much as a factor of two from the minority's proportion in the electorate should one want to motivate district races with the strategyproofness constraint.

Consider what might have happened under ADV after the June 1983 British general election. The Alliance (comprising the new Social Democratic party and the old Liberal party) got more than 25 percent of the national vote but less than 4 percent of the seats in Parliament (by coming in second in most districts); ADV would have assigned the Alliance 20 percent representation in an augmented 817-seat Parliament—26 percent larger than the present 650 seats.[36] Of course, under ADV it might be advisable to cut considerably the number of districts in Britain so that, after the usual adjustments are made, Parliament will be close to its present size and not too unwieldy.

A number of practical as well as theoretical issues are raised by ADV. For example, how should one deal with ethnic, linguistic, religious, racial, or other groups that are not represented by a single political party and whose size can therefore not be determined from the electoral returns? Will uncertainty about the final size of the legislature create problems?

In most parliamentary democracies "list systems," which allow voters to vote only for national parties rather than candidates in their own districts, are used (candidates are chosen from party lists to approximate the national vote division). Perhaps the main advantage of ADV over list systems is that voters, not parties, make the final selection of candidates, although the candidates who run in districts are normally affiliated with one of the national parties. Also, under ADV, voters have their own representative; moreover, if it is stipulated that this person must live in the district he or she represents, presumably that representative is even more accountable to the voters. True, voters can vote directly for candidates under the Hare system, but ADV approximates PR without the nonmonotonicity of Hare.

4.9 The Apportionment Problem

As attempts to achieve PR are confounded by strategyproofness under ADV, so are there fundamental conflicts among rules for apportioning seats to states or to parties to achieve fair representation.

The apportionment problem is best illustrated by the following concrete question: How can one assign seats to states in the U.S. House of Representatives to conform with the constitutional edict that "representatives shall be apportioned ... according to [the states'] respective numbers" (Art. 1, sec. 2).[37] The same problem arises in apportioning parliamentary seats to political parties in proportion to their popular vote or instructors to university departments in proportion to departmental enrollments.

The apportionment problem arises from the fact that—unlike taxes, which the Constitution also prescribes must be equitably apportioned among the states—representatives cannot be fractionalized to yield a perfect alignment of state populations and state representation. Integer assignments are further complicated by the constitutional stipulation that each state is entitled to at least one representative.

How to handle the fractions fairly was—and continues to be—the subject of major disagreement. Different solutions were proposed by, among others, Alexander Hamilton, Thomas Jefferson, and Daniel Webster. George Washington favored Secretary of State Jefferson's method over Secretary of the Treasury Hamilton's, casting the first presidential veto aginst the latter's bill in 1792. Webster's method was adopted following the 1840 census, after which a switch was made to Hamilton's method; the Webster method was resurrected after the 1900 census, and still another method is used today.

Hamilton's method is the simplest conceptually. One first computes the exact number of seats, fractions included, to which each state is entitled at some time (two times, t and t + 1, will later be compared for the hypothetical legislature shown in Table 4.4), by dividing the average district size into each state's population. (In Table 4.4, the average district size at t and t + 1 is the total population, 1,050, divided by the total number of seats to be apportioned, 7, or 150.) This is called the state's *quota* (for example, the quota for state 1 at t + 1 is 752/150 = 5.013). Next, one assigns to each state the integer part of its quota. Seats that are left over are then assigned to states whose quotas have the largest fractional parts. Thus, state 1 gets five seats at time t + 1, and the remaining two seats go to states 2 and 3, which have the two largest fractional parts in their quota.

In contrast to the Hamilton method, the Webster method focuses on the population of a representative's district. The goal is to choose a size such that the number of seats to which each state is entitled, when rounded to the nearest integer, sums to the size of the body. Thus, in the example in Table 4.4 at t + 1, if one sets the district size at 170, states 1 through 4 will be entitled to 4.424, 0.594, 0.582, and 0.576 seats, respectively. (Divisors other than 170, such as 169 and 171, will do as well; divisors that lead to a body of specified size will in general be different from the average district size under Hamilton's method.) These figures, which round to 4, 1, 1, and 1, respectively, sum to the desired legislative size of 7, whereas the average district size of 150 gives quotas that round to 8 at t and t + 1. Jefferson's method, and a method championed by the

Table 4.4 Apportionments of Webster and Hamilton Methods

	Time t				Time t + 1			
State	Population	Quota	Webster	Hamilton	Population	Quota	Webster	Hamilton
1	598	3.987	4	4	752	5.013	4	5
2	299	1.993	2	2	101	0.673	1	1
3	76	0.507	0	0	99	0.660	1	1
4	77	0.513	1	1	98	0.653	1	0
Total	1,050	7.000	7	7	1,050	7.000	7	7

Source: Steven J. Brams and Philip D. Straffin, Jr., "The Apportionment Problem," *Science* 217 (July 30, 1982); 437-438, Table 1. Reprinted by permission.

mathematician E. V. Huntington that is in use today, are based on the same idea but differ in how they do the rounding, with the former tending to favor larger states, the latter smaller states.

Reasonable as these apportionment methods may seem, they can produce anomalous results, as illustrated by the apportionments in Table 4.4 at the two different times t and t + 1.

1. *The Webster method violates quota.* If a state's quota is 5.013 (state 1 at t + 1), it seems sensible that an apportionment method should give it this number either rounded down or rounded up (5 or 6 seats); yet Webster at t + 1 gives state 1 only 4 seats.
2. *The Hamilton method violates population monotonicity.* Between t and t + 1, state 4's population increases by 27 percent (from 77 to 98), but it loses a seat, whereas state 1, whose population increases by only 26 percent (from 598 to 752), gains a seat.

The Hamilton method is also subject to other difficulties, such as the "Alabama paradox," whereby *increasing* the legislature size can *decrease* the representation of some states. Alabama was threatened with such a decrease after the 1880 census, but this does not seem to be a problem today because, since 1912, the actual size of the House of Representatives has remained fixed at 435 members (except for a temporary increase to 437 when Alaska and Hawaii were admitted as states in 1959).

Can one find apportionment methods that do not behave in these paradoxical ways? Balinski and Young prove that the answer is no: there is no apportionment method that both always satisfies quota and is monotonic in population. In fact, the general argument is already implicit in the examples in Table 4.4. The only apportionments that satisfy quota and do not give larger states fewer seats than smaller states are (4,2,0,1), (4,1,1,1), and (3,2,1,1,) at t and (5,1,1,0) and (6,1,0,0) at t + 1. If one is to satisfy quota, state 4 *must* lose its seat at t + 1.

Balinski and Young opt for the population-monotonicity condition over the quota condition, strongly arguing for Webster as the preferred apportionment method. They demonstrate the Webster method to be uniquely free of bias against small or large states and much less liable to violate quota than any other methods that satisfy population monotonicity.

Yet violations of quota are salient and could be politically disturbing. By comparison, population monotonicity is quite subtle. Balinski and Young claim that under a method that does not satisfy this kind of population-apportionment consistency over time "a state could deliberately undercount its population or encourage emigration to obtain an increase in its representation." [38]

This claim is misleading, however. Not only would the state have to fabricate new lower figures for itself but it would also have to arrange for crucial miscounts in other states as well. (Without carefully constructed changes in other states' figures, a simple drop in a state's population can never raise its apportionment under a nonmonotonic method like Hamilton's.) Apart from the preposterousness of such subterfuges, it is not obvious in the Table 4.4 examples that state 4's apportionment and population at t (say, in 1970) should have a necessary bearing on its apportionment and population at t + 1 (in 1980), and for 10 years thereafter (until 1990), by giving it less than its quota rounded down.

Is satisfying population monotonicity (Webster) more important that guaranteeing quota (Hamilton), especially when quota violations seem rare under the Webster method? The debate on this question is likely to continue. Indeed, when it was discovered that if Webster were substituted for the current method one seat would shift from New Mexico to Indiana on the basis of the 1980 census, not surprisingly the Indiana House delegation sponsored a reform bill.[39]

There is no perfect method of apportionment, but it is important to clarify the principles of fair representation and show the fundamental conflict among some of these principles. An understanding of the nature of the conflict may not lead to agreement but surely will make debate about the best method of apportionment more informed.

4.10 Conclusions

Paradoxes are ubiquitous in the social-choice literature, but the paradox of voting is probably the most famous (or infamous). As manifested in the phenomenon of cyclical majorities, it may crop up either in direct voting on alternatives or in indirect (platform) voting on parties and candidates.

More generally, the existence of the paradox does not depend on any particular voting procedure or decision rule. *Any* method of summing connected and transitive individual preferences that satisfies the five conditions of Arrow's general possiblity theorem is vulnerable to the paradox. Does the existence of the paradox therefore doom any social-choice process to an inescapable arbitrariness?

In one sense it does, because there is no procedure or type of voting so far considered that can circumvent the arbitrariness of the paradox. Indeed, the arbitrary social choices that may be foisted on members of a society because of the paradox have alarmed some analysts. Others have argued that this consequence is of little moment.[40] Although the latter judgment is difficult to accept in light of both the probabilistic calculations and—more important—the empirical instances of the paradox discussed, various of Arrow's conditions, especially universal admissibility of individual preference scales, are probably violated sufficiently often in practice that the general possiblity theorem cannot be considered an impenetrable barrier to coherent social processes.

In any event, it is probably foolish to expect that the standards by which we judge individual choices be appropriate to judgments about social choices. As Riker and Ordeshook put it:

> We are not deeply disturbed by the paradox, for it serves mainly to remind us that society is not the same as the people who compose it. People are not invariably disturbed by the inconsistencies and incoherencies of market outcomes—such as the oft-discovered fact that society spends more on liquor than education though surely a majority would wish otherwise. Markets have been churning out such inconsistencies for centuries without leading us to reject them as useful tools. Similarly there is no reason to reject other institutions of summation simply because they also are incoherent by human (i.e., individual) standards.[41]

Thus, the most important lesson of the paradox of voting may be not only that there is a qualitative difference between individual and social choice but also that one should not expect otherwise.

As for the monotonicity paradox, it seems antithetical to democratic choice that raising a candidate in a voter's preference scale should hurt him or her. Runoff and elimination systems, such as Hare, that are nonmonotonic would therefore seem seriously flawed, though reactions of voters to poll information, as was shown, may render other systems also vulnerable to nonmonotonicity.[42]

Analogously, representation presents irreconcilable conflicts. I suggested a variable-size legislature as a way to ameliorate certain problems with extant proportional representation systems; unfortunately, under such a system full PR is not consistent with measures taken to rob candidates of the incentive to lose district elections. Similarly, there are major conflicts in apportioning representatives to states in a completely fair way.

These paradoxes and conflicts underscore the difficulty of designing voting and representation systems that possess all the most desirable properties. Understanding the difficulties, as well as how certain trade-offs can be made among desirable properties, facilitates the rational design of institutions to achieve ends one thinks most important. To be sure, rational constructions must be matched against empirical experience to spotlight flaws that might not be apparent in theory. Once these theoretical and empirical strands are interwoven into a fabric describing what is possible, then normative judgments about what is desirable can be more rigorously made.

NOTES

1. In this case, a voter's distance from a candidate's position must be measured in two- or higher-dimensional space, making optimal positions of candidates with respect to different distributions of voter attitudes considerably harder to analyze theoretically and measure statistically. See James M. Enelow and Melvin J. Hinich, *The Spatial Theory of Voting: An Introduction* (New York: Cambridge University Press, 1984).
2. Claude Hillinger, "Voting on Issues and on Platforms," *Behavioral Science* 16 (November 1971): 564-566; see also Joseph B. Kadane, "On Division of the Question," *Public Choice* 13 (Fall 1972): 47-54, for an analysis of the effects of combining different alternatives.
3. Anthony Downs, *An Economic Theory of Democracy* (New York: Harper and Row, 1957), chap. 4.
4. Hillinger, "Voting on Issues and on Platforms," p. 565, claims this is not the case, but he is refuted in Nicholas R. Miller, "Logrolling and the Arrow Paradox: A Note," *Public Choice,* 21 (Spring 1975): 110.
5. Norman Frohlich and Joe A. Oppenheimer, *Modern Political Economy* (Englewood Cliffs, N.J.: Prentice-Hall, 1978), p. 135.
6. Downs, *Economic Theory of Democracy,* chap. 8.
7. William H. Riker and Peter S. Ordeshook, *An Introduction to Positive Political Theory* (Englewood Cliffs, N.J.: Prentice-Hall, 1973), pp. 114-115.
8. Kenneth J. Arrow, *Social Choice and Individual Values,* 2d ed. (New Haven, Conn.: Yale University Press, 1963); the first edition of this classic work was published in 1951.
9. Riker and Ordeshook, *Introduction to Positive Political Theory,* p. 91.
10. For reviews and syntheses of the social-choice literature related to the general possibility theorem, see Amartya K. Sen, *Collective Choice and Social Welfare* (San Francisco: Holden-Day, 1970); Prasanta K. Pattanaik, *Voting and Collective Choice: Some Aspects of the Theory of Collective Decision Making* (New York: Cambridge University Press, 1971); Peter C. Fishburn, *The Theory of Social Choice* (Princeton, N.J.: Princeton University Press, 1973); David J. Mayston, *The Idea of Social Choice* (New York: St. Martin's, 1974); Jerry S. Kelly, *Arrow Impossibility Theorems* (New York: Academic, 1978); Dennis C. Mueller, *Public Choice* (Cambridge: Cambridge University Press, 1979); Alfred F. MacKay, *Arrow's Theorem: A Case Study in the Philosophy of Economics* (New Haven, Conn.: Yale University Press, 1980); Robert Abrams, *Foundations of Political Analysis: An Introduction to the Theory of Collective Choice* (New York: Columbia University Press, 1980); and Allan M. Feldman, *Welfare Economics and Social Choice Theory* (Boston: Martinus Nijhoff, 1980).
11. Condition 3 does not say that the set of alternatives can change (for example, by the elimination of an "irrelevant" alternative), as suggested by the example in Arrow, *Social Choice and Individual Values,* p. 27. What the condition says is that *preferences* with respect to an irrelevant alternative (z in the example) can change; on this point, see Charles R. Plott, "Recent Results in the Theory of Voting," in *Frontiers of Quantitative Economics,* ed.

M. D. Intriligator (Amsterdam: North-Holland, 1971), pp. 109-129; and Plott, "Ethics, Social Choice Theory and the Theory of Economic Policy," *Journal of Mathematical Sociology* 2 (July 1972): 181-208.

12. Black's results are included in Duncan Black, *Theory of Committees and Elections* (Cambridge: Cambridge University Press, 1958).
13. Amartya K. Sen, "A Possibility Theorem on Majority Decisions," *Econometrica* 34 (April 1966): 491-499. For other characterizations of partial agreement that ensure a transiive social preference, see Sen, *Collective Choice and Social Welfare,* pp. 169-171.
14. Riker and Ordeshook, *Introduction to Positive Political Theory,* p. 97.
15. William H. Riker, "Arrow's Theorem and Some Examples of the Paradox of Voting," in *Mathematical Applications in Political Science,* ed. John M. Claunch (Dallas: Arnold Foundation of Southern Methodist University, 1965), pp. 41-60.
16. Riker, it should be pointed out, uses the word "contrivance" to refer not to insincere voting but only to the device of introducing a third alternative that creates a paradox-of-voting situation whereby a previous Condorcet candidate (O in the example) is rendered vulnerable to the new alternative (A in the example). (Alternative A is in turn vulnerable to the third alternative, N, which is itself vulnerable to O, thereby completing the cycle.)
17. When the possibility of strategic voting under his method of vote counting (described in Section 4.4) was pointed out to Jean-Charles de Borda (1733-1799), he replied, "My scheme is only intended for honest men." Black, *Theory of Committees and Elections,* pp. 183, 238. Besides the fact that the Borda count violates Arrow's independence condition, it is extremely vulnerable to strategic manipulation. See, for example, Philip D. Straffin, Jr., *Topics in the Theory of Voting* (Boston: Birkhäuser, 1980), pp. 27-29, and references cited therein. Work by Gibbard and Satterthwaite related to the general possibility theorem indicates that any scheme that is invulnerable to manipulation is necessarily dictatorial. See Alan Gibbard, "Manipulation of Voting Schemes: A General Result," *Econometrica* 41 (May 1973): 587-601; and Mark Allan Satterthwaite, "Strategy Proofness and Arrow's Conditions: Existence and Correspondence Theorems for Voting Procedures and Social Welfare Functions," *Journal of Economic Theory* 10 (April 1975): 197-218.
18. Robin Farquharson, *Theory of Voting* (New Haven, Conn.: Yale University Press, 1969), pp. 52-53. Farquharson offers a general game-theoretic analysis of what he calls "sophisticated" (strategic) voting; in Chapter 5 I shall analyze sophisticated (and deceptive) voting more formally.
19. *Congressional Quarterly Almanac,* vol. 11, 84th Cong., 1st sess. (Washington, D.C.: Congressional Quarterly Inc., 1955), p. 438.
20. Whereas contrivance of the paradox normally involves introduction by amendment of a third alternative, here an amendment was used in effect to suppress an alternative.
21. Farquharson, *Theory of Voting,* p. 53.
22. For accounts of paradoxes, some apparently contrived, that involved more than three alternatives or voters, see the following: William H. Riker, in "The

Paradox of Voting and Congressional Rules for Voting on Amendments," *American Political Science Review* 52 (June 1958): 349-366, analyzes voting on amendments to an appropriations bill for the Soil Conservation Service in the U.S. House of Representatives in 1952; Michael J. Taylor, in "Graph-Theoretical Approaches to the Theory of Social Choice," *Public Choice* 4 (Spring 1968): 45-46, analyzes voting in a committee of a university department; Richard G. Niemi, in "The Occurrence of the Paradox of Voting in University Elections," *Public Choice* 8 (Spring 1970): 91-100, analyzes voting in several elections held by a university faculty; and John C. Blydenburgh, in "The Closed Rule and the Paradox of Voting," *Journal of Politics* 33 (February 1971): 57-71, analyzes voting in the U.S. House of Representatives on amendments to the Revenue Acts of 1932 and 1938. The most systematic investigations of the paradox in the Senate and House are Bruce D. Bowen, "Toward an Estimate of the Frequency of the Paradox of Voting in U.S. Senate Roll Call Votes," and Herbert F. Weisberg and Richard G. Niemi, "Probability Calculations for Cyclical Majorities in Congressional Voting," both in *Probability Models of Collective Decision Making,* ed. Richard G. Niemi and Herbert F. Weisberg (Columbus, Ohio: Charles E. Merrill, 1972), pp. 181-203 and 104-131.

23. For examples, see Steven J. Brams, *Game Theory and Politics* (New York: Free Press, 1975), p. 92; and Brams, *Paradoxes in Politics: An Introduction to the Nonobvious in Political Science* (New York: Free Press, 1976), p. 48.
24. Black advances a similar argument that the later an alternative is introduced into the voting, the more likely it will be adopted (by a majority, but not necessarily by a coalition whose members have agreed to cooperate). Black, *Theory of Committees and Elections,* pp. 39-45. But Farquharson shows that although this is true when voting is sincere, if voting is sophisticated an alternative fares better the earlier it is introduced. Farquharson, *Theory of Voting,* p. 62. See also Bernard Grofman, "Some Notes on Voting Schemes and the Will of the Majority," *Public Choice* 7 (Fall 1969): 65-80.
25. Material in the remainder of this section is drawn from Steven J. Brams and Peter C. Fishburn, *Approval Voting* (Boston: Birkhäuser, 1983), pp. 141-145.
26. Gideon Doron and Richard Kronick, "Single Transferable Vote: An Example of a Perverse Social Choice Function," *American Journal of Political Science* 21 (May 1977): 303-311.
27. Suspicions seemed to have been first voiced in *Report of the Royal Commission Appointed to Enquire into Electoral Systems* (London: HMSO, 1910), Cd. 5163, p. 21; and James Creed Meredith, *Proportional Representation in Ireland* (Dublin, 1913), p. 93. I am grateful to Duff Spafford for these references. Confirmation of the paradox came in John H. Smith, "Aggregation of Preferences with Variable Electorate," *Econometrica* 42 (November 1973): 1027-1041.
28. Peter C. Fishburn, "Monotonicity Paradoxes in the Theory of Elections," *Discrete Applied Mathematics* 4 (April 1982): 119-134. A related paradox, in which some voters—by truncating their preferences (for example, indicating only a first choice rather than an ordering over all candidates)—can induce a

better outcome under the Hare system, is given in Steven J. Brams, "The AMS Nomination Procedure Is Vulnerable to 'Truncation of Preferences,' " *Notices of the American Mathematical Society* 29 (February 1982): 136-138, where an extreme case of nonmonotonicity is also given: a candidate loses when some voters raise him or her from last to first place in their preference scales. See also Steven J. Brams and Peter C. Fishburn, "Some Logical Defects of the Single Transferable Vote," in *Choosing an Electoral System: Issues and Alternatives,* ed. Bernard Grofman and Arend Lijphart (New York: Praeger, 1984), pp. 147-151. The truncation problem is also shown to affect other preferential voting systems in Hannu Nurmi, "On Taking Preferences Seriously," in *Essays on Democratic Theory,* ed. Dag Anckar and Erkki Berndtson (Tampere, Finland: Finnish Political Science Association, 1984), pp. 81-104, and Nurmi, "On the Strategic Properties of Some Modern Methods of Group Decision Making," *Behavioral Science* 29 (October 1984): 248-257; this problem is generalized by Peter C. Fishburn and Steven J. Brams, "Manipulability of Voting by Sincere Truncation of Preferences," *Public Choice* 44 (1984): 397-410, where its dependence on certain formal conditions is specified. For a catalog of paradoxes to which preferential voting is vulnerable, see Peter D. Fishburn and Steven J. Brams, "Paradoxes of Preferential Voting," *Mathematics Magazine* 56 (September 1983): 207-214.

29. A detailed analysis of this voting system, and comparisons with other systems, is given in Brams and Fishburn, *Approval Voting.*
30. Steven J. Brams, "Strategic Information and Voting Behavior," *Society* 19 (September/October 1982): 4-11; and Brams and Fishburn, *Approval Voting,* chap. 7.
31. A precise rule for determining the quota, and paradoxes to which the Hare system may give rise, are given in Fishburn and Brams, "Paradoxes of Preferential Voting."
32. The remainder of this section draws on Steven J. Brams and Peter C. Fishburn, "A Note on Variable-Size Legislatures to Achieve PR," in *Choosing an Electoral System: Issues and Alternatives,* ed. Bernard Grofman and Arend Lijphart (New York: Praeger, 1984), pp. 175-177. For an excellent discussion of the meaning of proportional representation in both theory and practice, and a new definition of representativeness with a procedure to implement it, see John R. Chamberlin and Paul R. Courant, "Representative Deliberations and Representative Decisions: Proportional Representation and the Borda Rule," *American Political Science Review* 77 (September 1983): 718-733.
33. Edward Still, "Alternatives to Single-Member Districts," in *Minority Vote Dilution,* ed. Chandler Davidson (Washington, D.C.: Howard University Press, 1984), pp. 249-267; and Hansard Society Commission, *The Report of the Hansard Society Commission on Electoral Reform* (London: Hansard Society for Parliamentary Government, 1976). Legal and constitutional arguments for the kind of PR system developed in this section are presented in John R. Low-Beer, "The Constitutional Imperative of Proportional Representation," *Yale Law Journal* 94 (November 1984): 163-188.

34. These and related voting systems are discussed in, among other places, Vernon Bogdanor, *What Is Proportional Representation? A Guide to the Issues* (Oxford: Martin Robertson, 1984); Vernon Bogdanor, *The People and the Party System: The Referendum and Electoral Reform in British Politics* (Cambridge: Cambridge University Press, 1981); Peter J. Taylor and Ronald J. Johnston, *Geography of Elections* (London: Penguin, 1979); Graham Gudgin and Peter J. Taylor, *Seats, Votes, and the Spatial Organization of Elections* (London: Pion, 1978); Enid Lakeman, *How Democracies Vote: A Study of Electoral Systems,* 4th rev. ed. (London: Faber and Faber, 1974); Arend Lijphart, *Democracies: Patterns of Majoritarian and Consensus Government in Twenty-One Countries* (New Haven, Conn.: Yale University Press, 1984); and Michael Dummett, *Voting Procedures* (Oxford: Oxford University Press, 1984).
35. Steven J. Brams and Peter C. Fishburn, "Proportional Representation in Variable-Size Legislatures," *Social Choice and Welfare* 1 (1984): 211-229.
36. Ibid.
37. This section draws on Steven J. Brams and Philip D. Straffin, Jr., "The Apportionment Problem," *Science,* July 30, 1982, pp. 437-438, which is a book review of Michel L. Balinski and H. Peyton Young, *Fair Representation: Meeting the Ideal of One Man, One Vote* (New Haven, Conn.: Yale University Press, 1982). The book offers a colorful history of apportionment along with an elegant development of the mathematical theory; Balinski and Young's main results will be described in this section.
38. Balinski and Young, *Fair Representation,* p. 68.
39. Alan Murray, "Reapportionment: The Politics of N(N − 1)," *Congressional Quarterly Weekly Report,* February 28, 1981, p. 393.
40. For a penetrating assessment of different views, see Riker and Ordeshook, *Introduction to Positive Political Theory,* chap. 6.
41. Ibid., p. 114. William H. Riker, in *Liberalism against Populism: A Confrontation between the Theory of Democracy and the Theory of Social Choice* (San Francisco: Freeman, 1982), compares two different notions of democracy and concludes that only "liberalism" (presuming some control over public officials through periodic elections), which is less demanding of voting processes than "populism" (presuming that the will of the people both can be expressed and is morally right), survives the assault of social choice theory.
42. Hannu Nurmi offers a good summary evaluation of common voting systems in "Voting Procedures: A Summary Analysis," *British Journal of Political Science* 13 (April 1983): 181-208; he concludes, on the basis of several criteria, that approval voting is the best of the lot. A similar conclusion is reached by Robert F. Bordley in "A Pragmatic Method for Evaluating Election Schemes through Simulation" *American Political Science Review* 72 (September 1983): 831-847; and by Samuel Merrill III in "Strategic Decisions under One-Stage Multi-Candidate Voting Systems," *Public Choice* 36 (1981): 115-134, and "A Comparison of Efficiency of Multicandidate Electoral Systems," *American Journal of Political Science* 28 (February 1984): 23-48. There has, however, been a good deal of controversy about the probable effects of

approval voting, some of which can be found in Theodore S. Arrington and Saul Brenner, "Another Look at Approval Voting," *Polity* 17 (September 1984): 118-134, and Steven J. Brams and Peter C. Fishburn, "A Careful Look at 'Another Look at Approval Voting,'" *Polity* 17 (September 1984): 135-143; and Richard G. Niemi, "The Problem of Strategic Voting under Approval Voting," *American Political Science Review* 78 (December 1984): 952-958, which provoked an exchange between Niemi, and Brams and Fishburn, in "Communications," *American Political Science Review* 79 (September 1985).

Voting Power 5

5.1 Introduction

Power is probably the most suggestive concept in the vocabulary of political scientists. It is also one of the most intractable concepts, bristling with apparently contradictory meanings and implications. For example, one implication of most definitions of power is that the greater the proportion of resources (such as votes) that an actor controls, the greater is his or her power. In this chapter this implication, under certain conditions, will be shown to be false.

I shall illustrate its falsity in different ways and then develop a general argument that the power of an actor, at least in voting bodies, can best be conceptualized as—and measured by—that actor's *control over outcomes*. To demonstrate that the resources that an actor possesses may be chimerical in achieving better outcomes, I shall show that a chair in a voting body, who it will be initially assumed can cast a tie-breaking vote in addition to a regular vote, may be at a disadvantage vis-à-vis the regular members when the preference rankings of outcomes by the members all differ. This phenomenon is called the "paradox of the chair's position," and possible ways that a chair may circumvent this paradox either through deception or under different voting rules will be discussed.

To render the paradoxical relationship between votes and power more precise, an index of voting power due to Banzhaf will be defined. The application of the Banzhaf index to both hypothetical and real voting bodies indicates that an actor's voting power may actually be greater when he or she controls a smaller, rather than a larger, proportion of the votes in a weighted voting body (one in which the members may cast different numbers of votes).

This and other anomalous results demonstrate that greater resources may not translate into greater ability to control outcomes. Such findings, however, do not establish the optimal *means* for getting one's way in different situations. For this purpose, it is convenient to postulate that players in voting games have preferences for outcomes and choose strategies consistent with getting their best possible outcomes, as was done in the case of the empirical examples of the paradox of voting in Section 4.6. More calculations of this sort will be illustrated in this chapter; in the next chapter, the effects of being powerful will be analyzed outside the rule-bound setting of voting bodies.

5.2 The Paradox of the Chair's Position

It would seem that the chair of a voting body would have more power than other members if it has a tie-breaking vote in addition to a regular vote. Yet, under circumstances to be described, a chair may actually be at a disadvantage relative to other members.

Consider the three-voter, three-alternative situation discussed in Section 4.2 that leads to a paradox of voting (Table 4.1) in which the preference scales of the voters all differ. Assume that the voting procedure is the plurality procedure, whereby the alternative with the most votes wins, and that the voter with preference scale (x,y,z) (called x-voter) has more votes than either of the other two voters but not enough to win if the other two voters vote for another alternative.

Under a different interpretation, x-voter may be considered to have the same number of votes as the other two voters but, in the event of a three-way tie, can act as the chair and break ties. In either event, it would appear that x-voter has some kind of edge over the other two voters, y-voter and z-voter, with preference scales (y,z,x) and (z,x,y).

If voting is sincere, x-voter will prevail since each voter will vote for the alternative he or she ranks first. By being able to make the decisive choice in the case of a three-way split, x-voter can ensure the selection of x.

Curiously, however, x-voter's apparent advantage in voting power over the other two voters disappears if voting is sophisticated (to be defined shortly). First, note that x-voter has a dominant strategy of "vote for x," which is never worse, and sometimes better, whatever the other two voters do (Section 2.2). For if the other two voters vote for the same alternative, x-voter cannot exercise a tie-breaking vote and so cannot improve the outcome. On the other hand, if the other two voters disagree, x-voter's tie-breaking vote will be decisive in the selection of x. Hence, x-voter's sincere vote is best whatever the circumstances.

Given this choice on the part of x-voter, y-voter and z-voter face the strategy choices shown in Figure 5.1. Their dominated strategies are crossed out, leaving for y-voter two undominated strategies, "vote for y" and "vote for z"; and for z-voter, one undominated strategy, "vote for z." In the case of y-voter, "vote for x," which always leads to his or her worst alternative, is dominated by both the other strategies. In the case of z-voter, "vote for z" is actually a dominant strategy, or a unique undominated strategy, because it is at least as good and sometimes better than his or her other two strategies, whatever strategy y-voter adopts.

If voters have complete information about each other's preferences, they can perceive the situation in terms of Figure 5.1 and eliminate the dominated strategies. Then y-voter, choosing between "vote for y" and

Figure 5.1 Outcomes under Plurality Procedure When X-Voter Always Chooses x

First reduction

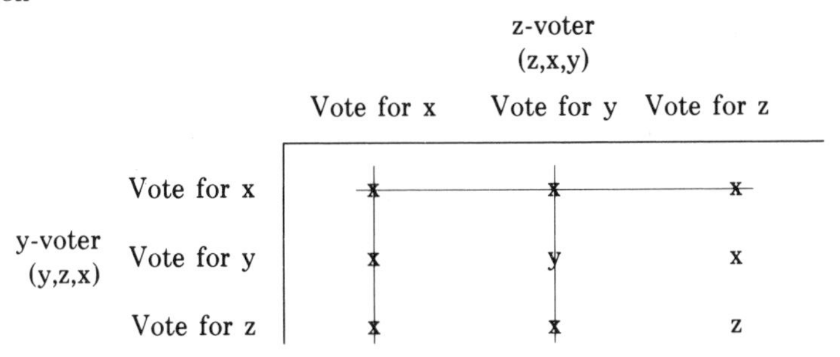

Second reduction

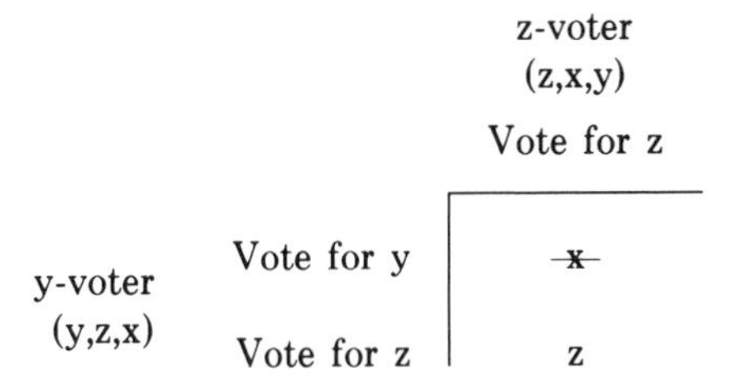

Note: The dominated strategies of each voter are crossed out in the first reduction, leaving two (undominated) strategies for y-voter and one (dominant) strategy for z-voter. Given these eliminations, y-voter would then eliminate "vote for y" in the second reduction, making "vote for z" the sophisticated strategies of both voters and z the sophisticated outcome.

"vote for z" in the second-reduction matrix, would cross out "vote for y," now dominated because that choice would result in x's winning due to the chair's tie-breaking vote; instead, y-voter would choose "vote for z," ensuring z's election.

In this manner z, which is not the choice of a majority and could in fact be beaten by y in a pairwise contest, becomes the *sophisticated outcome.* In general, the successive elimination of dominated strategies by voters, insofar as this is possible, is called *sophisticated voting,* and the strategies that remain after all eliminations can be made are called *sophisticated strategies.*[1]

In game-theoretic terms, sophisticated voting produces a different and smaller game in which some formerly undominated strategies in the larger game become dominated in the smaller game. The removal of such strategies—sometimes in several successive stages—in effect enables sophisticated voters to determine what outcomes eventually *will* be adopted by eliminating those outcomes that definitely *will not* be chosen. Thereby voters attempt to foreclose the possibility that their worst outcomes will be chosen by the successive removal of dominated strategies, given the presumption that other voters do likewise.

How does sophisticated voting affect the chair's presumed extra voting power? Observe that the chair's tie-breaking vote is not only not helpful but positively harmful: it guarantees that x-voter's worst outcome will be chosen if voting is sophisticated! The illusory nature of the chair's extra voting power in such situations is called the *paradox of the chair's position.*

If this result seems anomalous, it is perhaps worth investigating how often the preference scales of the three voters will be such that the chair, despite possessing a tie-breaking vote, will not get its first choice if voting is sophisticated. In situations in which two or all three voters rank one alternative best, this alternative will obviously be chosen under both sincere and sophisticated voting. The situations that are of interest, therefore, are those in which all three voters have different first preferences, which are *conflict situations.*

Recall from Section 4.6 that there are $6^3 = 216$ distinct ways in which three voters can order three alternatives. Now once the chair has chosen an ordering, of which there are six possibilities, there are four orderings that give the second voter a first choice different from the chair's, and two orderings that give the third voter a first choice different from both the chair's and the first voter's. Hence, there are $(6)(4)(2) = 48$ distinct conflict situations.

It turns out that besides the 12 paradox-of-voting situations (see Section 4.5), 24 of the remaining 36 conflict situations are vulnerable to the paradox of the chair's position, though not in such a drastic form as the paradox-of-voting situations that ensure the chair's worst choice under sophisticated voting.[2] In these situations, either the chair's second choice is chosen or the result is indeterminate (there is more than one sophisticated outcome). In none of these cases, however, can the chair's tie-breaking vote induce its first choice if voting is sophisticated.

Given this unfortunate state of affairs for a chair, it is reasonable to ask whether a chair (or larger voting faction), with apparently greater voting power, has any recourse in such situations. Otherwise, the strategic calculations of sophisticated voting would appear inevitably to nullify the chair's edge over the other voters.

Consider the paradox-of-voting situation given at the beginning of this section. The sophisticated outcome is z, which is supported by both y-voter and z-voter. Clearly, x-voter has no voting strategy that will alter the sophisticated outcome, x-voter's worst.

This result also holds in the other conflict situations in which the sophisticated outcome is either indeterminate or the second choice of the chair. Such a consequence would appear to be pretty dismaying to a chair: in 36 of the 48 cases in which the most-preferred outcomes of all three voters differ, sophisticated voting under the plurality procedure undermines the chair's apparent power advantage—specifically, its ability to obtain a most-preferred outcome if voting is sincere. Moreover, there appears to be no voting strategy that can restore it. Does any escape remain for a chair once sincerity in voting is lost?

5.3 The Chair's Counterstrategy of Deception

A chair is often in the unique position, after the other voters have already committed themselves, of being the last voter to have to make a strategy choice. Yet this position does not furnish a ready solution to the chair's problem if voting is truly sophisticated, for sophisticated voting implies that voters act upon both their own preferences and a knowledge of the preferences of the other voters. Therefore, the order of voting is immaterial: all voters can predict sophisticated choices beforehand and act accordingly. Even a chair's (unexpected) deviation from a sophisticated strategy under the plurality procedure cannot generally effect for it a more favorable outcome.

Assume for purposes of the subsequent analysis that the chair, by virtue of its position, can obtain information about the preference scales of the other two voters but they cannot obtain information about its preference scale. Assume further that each of the two regular members is informed of the other's preference scale. If voting is to be sophisticated, the chair's preference scale must be made known to the regular members; however, the chair is not compelled to tell the truth. The question is: Can a chair, by announcing a preference scale different from its true perference scale, induce a more-preferred sophisticated outcome?

Given that voting is sophisticated, the chair, having the tie-breaking vote, will always have a dominant strategy: vote for its most-preferred outcome. Thus, the other voters need only know the chair's announced first choice, and not its complete preference scale, to determine what its strategy choice will be.

Define a *deception strategy* on the part of the chair to be any *announced* most-preferred outcome that differs from its *honestly* most-preferred outcome. Call the use of a deception strategy by the chair *tacit deception,* since the other members, not knowing its honest preference scale, are not able to determine whether the chair's announcement is an honest representation of its most-preferred outcome. Tacit deception will be profitable for the chair if it induces a more-preferred social choice than does an honest representation of its preferences, given sophisticated voting by the other voters.

As an illustration, consider the paradox-of-voting situation discussed in Section 5.2. The chair (x-voter), by announcing its first choice to be outcome y rather than x, can induce the sophisticated outcome y, which it would prefer to z. This can be seen by substituting y's for all the x's in the Figure 5.1 first-reduction matrix, which will be the *perceived* outcomes by y-voter and z-voter after x-voter's deceptive announcement. By eliminating dominated strategies, as earlier, it is easy to show that the sophisticated outcome is y—as the two deceived voters perceive it. Thus, tacit deception, by changing the outcome from z to y, is profitable for x-voter.

Assume now that x-voter as chair actually votes for outcome x after announcing its (dishonest) preference for outcome y. Then the (manipulated) sophisticated outcome will be x, the chair's first preference. In other words, the chair can induce its preferred outcome by announcing a bogus preference scale and, contrary to the announcement, voting honestly in the end.

Call this kind of deception, which involves not only announcing a deception strategy but taking *deceptive action* as well, *revealed deception.* This deception is revealed in the voting process and is clearly more profitable for x-voter than tacit deception in a paradox-of-voting situation.

In general, a dishonest announcement by the chair may improve its position somewhat, as in the paradox-of-voting situation just illustrated, wherein a dishonest announcement ensures passage of the chair's next-best outcome. Such an announcement, however, followed by the chair's vote for its most-preferred outcome (thereby flouting this announcement) is still better: it ensures the chair's best outcome (x in the example) in the

36 conflict situations in which there is a paradox of the chair's position. Of course, revealed deception becomes apparent after the vote—unless the vote is secret—and probably cannot be used very frequently. If it were, the chair's announcements would quickly become unbelievable and lose their inducement value.

The deception strategy game I have sketched for the chair can naturally be played by a regular member if he or she is privy to information that the chair and the other regular member are not. (The results will not duplicate those for the chair, however, because the chair has an extra resource—the tie-breaking vote.) I shall not carry this analysis further, though, because my main purpose has been to demonstrate that there is a resolution (of sorts) to the paradox of the chair's position.[3] It requires, however, that the information available to some players in the game be restricted, which has the effect of endowing one player (the chair) with still greater resources.

This, it must be admitted, is itself a rather deceptive way out of a problem that seems genuine. If voting is sophisticated, or if coalitions can form, the chair, despite the added weight of its position, will not necessarily exercise greater control over outcomes than the other members. In fact, the reverse might be the case, as the paradox-of-voting situations under the plurality procedure illustrate.

Strangely enough, in these situations a chair can obtain its best outcome if the voting system used is approval voting (Section 4.7). Under this system, each voter would choose between two undominated strategies: vote for his or her best, or two best, choices.[4] The sophisticated strategies of the three voters turn out to be $\{x,y\}$—vote for both x and y—for x-voter (the chair), $\{y,z\}$ for y-voter, and $\{x,z\}$ for z-voter, which yields x when the chair breaks a three-way tie in favor of x. However, the chair may also do worse under approval voting than plurality voting if the preferences of the voters are different, so approval voting provides no panacea for a chair who suffers vis-à-vis the other voters solely because of the extra resources that it commands.

Two similar paradoxes may arise under either voting system, whereby a chair does worse (1) having both a regular and tie-breaking vote versus having only a tie-breaking vote (no-vote paradox); or (2) having a tie-breaking vote and participating versus not participating in the election at all (no-show paradox).[5] These consequences of sophisticated voting seem paradoxical because a chair in each case has ostensibly greater power (a tie-breaking vote in the no-show paradox, a regular vote in the no-vote paradox) that it would do better without or not using. Moreover, because the encumbrances caused by having extra resources cannot be foreclosed under either plurality or approval voting, the paradoxes are not a peculiarity of the voting system used.

All these paradoxes make clear that power defined as control over outcomes is not synonymous with power defined in terms of control over resources, at least those that a chair may possess.[6] The strategic situation facing voters intervenes and may cause them to reassess their strategies in light of the additional resources that a chair possesses. In so doing, they may be led to gang up against the chair—that is, to vote in such a way as to undermine the impact of these additional resources, handing the chair a worse outcome than it would have achieved without them. These resources in effect become a burden to bear, not power to relish.

To derive greater insight into other anomalous aspects of voting power, I shall next explore a related paradox in terms of a quantitative measure of voting power. This measure of power is not tied to the preferences of actors or the strategic choices implied by these preferences in a game. Rather, it is intended to provide an assessment of the power of actors based on the formal rules that interrelate them in a voting body.

5.4 Banzhaf Voting Power

The paradox of the chair's position suggests that a quantitative definition of voting power should incorporate the idea of control over outcomes. As I shall show, such control is not in general perfectly correlated with the number of votes one casts, just as the ability to break ties is not invariably helpful to a chair. Rather, control over outcomes also depends on how frequently, on the average, one can pool one's votes with those of others to form coalitions (analyzed extensively in Chapter 8) that can ensure an outcome favorable to oneself.

Although other definitions of influence and power stress the effects that actors can have on each other, for the purpose of defining the power of players in voting games an outcome-oriented measure is preferable to an actor-oriented measure.[7] In large voting bodies or even in the electorate, wherein the influence of each person on every other person is for all practical purposes negligible, an actor-oriented measure would suggest that no one has any power. In fact, if each person has one vote, each person has an equal chance to influence the outcome, which seems a more reasonable way to view power in voting situations.[8]

This view is not compatible with defining the voting power of an actor to be proportional to the number of votes he or she casts, because votes per se may have little or no bearing on outcomes. For example, in a three-member voting body {a,b,c}, in which a has 4 votes, b 2 votes, and c 1 vote, members b and c are powerless if the decision rule is simple majority (4 out of 7). Since the fact that members b and c together control 3/7 of the votes is irrelevant to the selection of outcomes by this body, call these members *dummies*. Member a is a *dictator*, on the other hand, by virtue of having enough votes alone to determine the outcome; all coalitions of which a is a member are necessarily winning. Note that a voting body can have only one dictator, whose existence renders all other members dummies, but there may be dummies and no dictator (as will be shown in Section 5.6).

The votes cast by a member of a voting body are relevant in the selection of outcomes only in the context of the number of votes cast by other members and the decision rule of the voting body. To illustrate a measure of voting power due to Banzhaf that utilizes this information,[9] consider a three-member voting body with member weights {3,2,2}, in which the decision rule is a simple majority of 4 out of 7 votes. Define the *voting power* of a member to be the *number of winning coalitions in which the member's defection from the coalition would render it losing*—which is a *critical defection—divided by the total number of critical defections for all members.* Denote a winning coalition, wherein the subtraction of at least one of its members would change its status from winning to losing, a *minimal winning coalition* (MWC). An MWC as defined here need not be vulnerable to the defection of *all* its members but must certainly be vulnerable to the defection of its largest member (see Section 5.5).

I shall distinguish the two 2-vote members in the example by the subscripts 1 and 2 (2_1 and 2_2). There are three distinct MWCs—($3,2_1$), ($3,2_2$), and ($2_1,2_2$)—whose members overlap but are not all identical with those of any other such coalition. Clearly, the subtraction of the 3-vote member from ($3,2_1$) and ($3,2_2$), the 2_1-vote member from ($3,2_1$) and (2_1, 2_2), and the 2_2-vote member from ($3,2_2$) and ($2_1,2_2$) would render each MWC losing. Altogether, therefore, there are six critical defections for the three members of the voting body.

Since each member's defection is critical in two MWCs, each member's proportion of voting power is $2/6 = 1/3$ by Banzhaf's definition. These fractional values for each member may conveniently be represented by the components of a power vector, which I call the *Banzhaf index* of a voting body. For the voting body {3,2,2} the Banzhaf index is (1/3,1/3,1/3) under simple majority rule, which indicates that the voting power of the one 3-vote member is the same as that of each of the two 2-vote members.

5.5 The Paradox of New Members

If a new 1-vote member is added to the voting body {3,2,2} so that it becomes {3,2,2,1}, how does the increase in the size of the body affect the voting power of the original members, given a decision rule of simple majority (now 5 out of 8)? In the enlarged voting body, there are six

MWCs: $(3,2_1)$, $(3,2_2)$, $(3,2_1,1)$, $(3,2_2,1)$, $(3,2_1,2_2)$, and $(2_1,2_2,1)$. The defection of the 3-vote member is critical in five coalitions, each of the 2-vote members in three, and the 1-vote member in one, making a total of 12 critical defections. The Banzhaf power values for the body $\{3,2,2,1\}$ are, therefore, (5/12,3/12,3/12,1/12) = (5/12,1/4,1/4,1/12). Note that a coalition of members [for example, $(3,2_1,2_2)$] may be minimal winning with respect to the defection of one member (3) but not with respect to the defection of other members (2_1 and 2_2). Recall that for a coalition to be considered minimal winning, it must be so with respect to the defection of at least one of its members but not necessarily of all.

While the power of the two 2-vote members decreases in the enlarged voting body, the 3-vote member surprisingly *increases* power in this body. Specifically, the 3-vote member's power increases from 1/3 = 0.33 to 5/12 = 0.42 by the Banzhaf index, despite the fact that its proportion of votes decreases from 3/7 = 0.43 in the original body to 3/8 = 0.375 in the enlarged body. On the basis of this redistribution of power caused by the addition of the 1-vote member to the original voting body, it seems reasonable to suppose that the 3-vote member would favor an *expansion* in the size of the voting body by one 1-vote member!

Lest one think that a member's greater power in the enlarged (versus the original) voting body is an artifact of a change in the decision rule (from 4 out of 7 in the original body to 5 out of 8 in the enlarged body), consider what the power of the three original members would be if they had operated under a decision rule of 5 out of 7, the same as that assumed in the enlarged body. Then the defection of each of the two 2-vote members (2_1 and 2_2) would be critical in one coalition apiece [$(3,2_1)$ and $(3,2_2)$], and the defection of the 3-vote member would be critical in both these coalitions as well as in the coalition $(3,2_1,2_2)$. Hence, the Banzhaf power values of the voting body $\{3,2,2\}$ under a decision rule of 5 out of 7 are (3/5,1/5,1/5).

Comparing these values with the corresponding Banzhaf power values in the enlarged voting body $\{3,2,2,1\}$, each of the two 2-vote members *increases* its voting power from 1/5 = 0.20 to 1/4 = 0.25. Thus, when the decision rule is the same in the original and enlarged bodies (5 votes), each of the two 2-vote members—rather than the 3-vote member—benefits from the addition of a 1-vote member to the original body.

The *decrease* in a member's proportion of votes in the enlarged voting body, simultaneous with the *increase* in its voting power, seems certainly paradoxical, especially since the new 1-vote member added to the voting body is not a dummy and by its presence deprives the other members together of some voting power. Under the decision rule of simple majority in the original and enlarged voting bodies, for example, the 1-vote member reduces the combined voting power of the three original members from a total of 1.00 to a total of 0.92. Despite this collective reduction in voting power of the three original members, however, the new member causes a redistribution in the share that remains so that the largest original member (3) benefits under simple majority rule. When the decision rule is the same (5 votes) in both voting bodies, the two smaller original members (2_1 and 2_2) benefit.

A *paradox of new members* occurs when one or more new members are added to a weighted voting body—with or without a change in the decision rule—and the voting power of one or more of the original members increases rather than decreases. Although other paradoxes connected with the measurement of voting power have been identified, the paradox of new members is probably the one of greatest empirical interest.[10]

5.6 Power Anomalies in the European Community Council of Ministers

When the European Community (EC) was expanded from six to nine members with the admission of England, Denmark, and Ireland in 1973, the voting weights of all the old members on the EC Council of Ministers, except Luxembourg, were increased by a factor of 2.5: France, Germany, and Italy went from 4 to 10 votes, and Belgium and the Netherlands from 2 to 5. Luxembourg, by contrast, only doubled its vote total from 1 to 2.[11]

The proportion of Banzhaf voting power of Luxembourg on the council, nevertheless, increased from 0 to 0.016. Despite the fact that the new council now included three new members, with a combined total of 16 votes (England, 10; Denmark and Ireland, 3 each), and Luxembourg's proportion of votes fell from $1/17 = 0.059$ in the old council to $2/58 = 0.034$ in the new council, it had greater voting power in the new council.

To understand, without detailed calculations, why the paradox of new members occurred in this case, consider Luxembourg's situation on the council between 1958 and 1973.[12] In acting on policy proposals of the European Commission, the Treaty of Rome (which established the European Economic Community, or Common Market, in 1958) required a qualified majority of 12 out of 17 votes for passage of a measure on this council. This meant that, for Luxembourg's vote to be critical, it had to be a member of an MWC with exactly 12 votes; its defection would then cause that MWC to become losing.

Unfortunately for Luxembourg, however, this event could never occur. Because the votes of the five other members were all even numbers (the three 4's of France, Germany, and Italy; the two 2's of Belgium and the Netherlands), an MWC with exactly 12 votes could never include Luxembourg's (odd) 1 vote. By comparision, there were several ways in which a 13-vote MWC could form that included Luxembourg, but then its defection would not render such an MWC losing, though the defection of any of the 2-vote or 4-vote members would.

This is why Luxembourg was a dummy on the pre-1973 council. Yet with the expansion of the council to nine members in 1973 and with the qualified majority needed for passage of commission proposals now set at 41 out of 58 votes, Luxembourg's 2 votes on the new council became critical in a small proportion (1.6 percent) of MWCs.[13]

The story does not end in 1973. With the admission of Greece to the EC in 1981, the weights of the nine members of the 1973 council were not altered, but Greece was assigned 5 votes, giving it the same weight as Belgium and the Netherlands. The previous qualified majority of 41 out of 58 votes for the 1973 council was raised to 45 out of 63 votes. Astonishingly, despite Luxembourg's further dilution of votes on the council (2/63 = 0.032), its Banzhaf voting power increased once again: from 0 (1958-1973) to 0.016 (1973-1981) to 0.041 (1981-1985), which is a 156 percent improvement over its previous (1973-1981) voting power.

Table 5.1 Banzhaf Voting Power of Members of the EC Council of Ministers, 1958-1985

	1958-1973		1973-1981		1981-1985	
Country	Weight	Power	Weight	Power	Weight	Power
France	4	0.238	10	0.167	10	0.158
Germany	4	0.238	10	0.167	10	0.158
Italy	4	0.238	10	0.167	10	0.158
Belgium	2	0.143	5	0.091	5	0.082
Netherlands	2	0.143	5	0.091	5	0.082
Luxembourg	1	<u>0</u>	2	<u>0.016</u>	2	<u>0.041</u>
England	—	—	10	0.167	10	0.158
Denmark	—	—	3	0.066	3	<u>0.041</u>
Ireland	—	—	3	0.066	3	<u>0.041</u>
Greece	—	—	—	—	5	0.082

Source: Steven J. Brams and Paul J. Affuso, "New Paradoxes of Voting Power on the EC Council of Ministers," *Electoral Studies* 4 (August 1985): 185. Reprinted by permission.

Note: Underscored values illustrate the paradox of new members (Luxembourg's voting power increases as its proportion of votes falls on the post-1973 and post-1981 councils) and the paradox that, on the post-1981 council, countries with different voting weights (Luxembourg versus Denmark and Ireland) may have exactly the same voting power. The decision rules (that is, qualified majorities) for the three councils are as follows: 12 out of 17 (1958-1973); 41 out of 58 (1973-1981); and 45 out of 63 (1981-1985).

In Table 5.1 the Banzhaf voting power values for all countries on the three councils are given. Of greater moment, perhaps, than the recurrence of the paradox of new members on the 1981 council is another major surprise: Luxembourg, with 2 votes, and Denmark and Ireland, the two next-largest countries with 3 votes each, have exactly the same voting power (0.041).

How could this happen? Consider the cases in which Denmark and Ireland will be critical but Luxembourg will not. These will occur when there is an MWC with 47 votes from which Luxembourg's defection will not be critical (its defection still leaves the MWC with a requisite 45 votes) but Denmark's or Ireland's will be (by reducing the MWC to losing status with only 44 votes).

An MWC with exactly 47 votes, however, is an impossibility. For, like the pre-1973 council, in which an MWC with exactly 12 votes could never include Luxembourg, an MWC on the post-1981 council with exactly 47 votes could never include Denmark or Ireland.

To demonstrate this, one must show that there can be no 44-vote coalition without the 3 votes of either Denmark or Ireland (or both). Such a coalition is an impossibility since no combination that includes a subset of the four 10-vote members, the three 5-vote members, one (but not both) of the 3-vote members, and the 2-vote member sums to precisely 44 votes. Hence, the 3 votes of Denmark or Ireland can never be critical in cases other than those in which the 2 votes of Luxembourg also are critical. Therefore, these three countries must all have the same voting power.

If this result is less anomalous than the paradox of new members, it may be more portentous politically. After all, the purpose, presumably, of assigning different weights to council members is to reflect differences in size, however roughly. (This reflection is indeed rough: Luxembourg, with a population of 0.36 million, is far smaller than both Denmark [5.12 million] and Ireland [3.40 million], though its voting weight on the council is only 33 percent less than that of these two countries.) But even this very imperfect proportionality is obliterated when one looks at voting power, which reveals absolutely no distinction between the two sets of countries, despite their differences in voting weight.

Significantly, if different qualified majorities had been chosen as the decision rule, a distinction could have been maintained. For example, if the rule were a qualified majority of 46 rather than 45 out of 63 votes, Luxembourg's Banzhaf voting power would have been 0.013, and Denmark and Ireland's 0.063, which much more closely reflects their population differences. Other decision rules, from a simple majority of 33 out of 65 to a rule of unanimity of 65 out of 65 (which gives all members a veto and, therefore, equal power), have different effects.

The choice of voting weights, as well as a decision rule, may be used to satisfy a variety of goals, including reflecting the different sizes of members, giving some members special prerogatives (for example, veto power), maximizing the absolute number of MWCs in which one or more members is critical, and so on. These need to be carefully considered and spelled out before the design of a legislative body like the council can be carried out. If they are not, and seemingly plausible but nevertheless arbitrary voting weights and decision rules are selected—without benefit of formal voting-power analysis—then the effects on the body's members may be not only unanticipated but also bizarre, as I have illustrated in the case of the council, which is the EC's highest decision-making body.

One might, of course, challenge the validity of a particular measure of voting power, like Banzhaf's, but the results discussed in this section hold true for other well-known measures, including that of Shapley and Shubik.[14] Moreover, a measure like Banzhaf's not only seems an eminently reasonable indicator of a crucial aspect of voting power—the ability of a member to change an outcome by changing its vote—but also highlights the fact that size (as reflected by voting weights) and voting power may bear little relationship to each other.

After a second occurrence of the paradox of new members in 1981, and an unprecedented failure of the council to register differences in voting power by an assignment of different voting weights, it seems evident that there is a need to exercise greater care in the design of institutions like the council. Indeed, the admission of Portugal and Spain to the EC in 1986, with voting weights of 5 and 8 votes, respectively, on the council, and a qualified-majority decision rule of 54 (out of 76) votes on the council, will lead to still another manifestation of the paradox of new members.[15]

In this new Community of Twelve, Denmark and Ireland, with 3 votes each, will increase their voting power from 0.041 to 0.046, or by 12 percent, illustrating the occurrence of the paradox of new members once again. All the other old members will lose power, Luxembourg most dramatically by plummeting to a Banzhaf value of 0.018, a decrease of 56 percent from its 1981-1985 value of 0.041. (The Banzhaf values of the 10-vote countries will fall from 0.158 to 0.129 and those of the 5-vote countries from 0.082 to 0.067; Spain's 8 votes will give it a Banzhaf value of 0.109.)

Given the possible recurrence of the paradox, voting rules and members' weights would seem worthy of more careful consideration, based on formal voting-power analysis, than they have heretofore been given. Thereby, unintended and anomalous effects of institutional change might be foreseen and avoided.

5.7 Conclusions

I have looked at the exercise of voting power under rather varied circumstances in this chapter. Nevertheless, each illustration of its use, both hypothetical and real, shares a common trait: *an actor is powerful through being able to affect the choice of an outcome in a desired way.*

The paradox of the chair's position reflects more the absence of power than its presence, because the chair's tie-breaking vote—or, equivalently, extra voting weight—is for naught. Indeed, this prerogative has an ironic effect in some situations: it ensures that the chair's last choice will be chosen under sophisticated voting.

Potential chairs, beware! One may maximize one's influence, or effect on the outcome, by relinquishing the chair, not seeking it out. Although the chair can improve its position through a policy of either tacit or revealed deception, the use of such a strategy depends on having information about the preference scales of other members that they do not have about the chair's. On the other hand, changing the voting system *may* work to the chair's advantage, as indicated in the case of approval voting, but such a switch offers no guarantee, for the paradox of the chair's position and related paradoxes seem quite independent of the voting system used.

These paradoxes probably extend to other political arenas. In international relations, for example, countries that steer clear of alliance involvements—as Austria and Switzerland have done in Western Europe—or pledge themselves not to develop nuclear weapons—as most countries in the world today have done by signing the Nuclear Nonproliferation Treaty—would seem examples in which shunning certain prerogatives or resources may redound to the advantage of the shunner. In fact, when a policy of apparent self-denial fosters superior collective choices for those doing the denying, it might better be characterized as a policy of self-enhancement. Under what strategic conditions the less powerful (in resources) become the more powerful (in securing better outcomes for themselves) is an intriguing question on which little research has been done.

The tenuous linkage between resources and control over outcomes is underscored by the analysis of voting power. As shown by the paradox of new members, the addition of one or more new members to a weighted voting body can increase the voting power of some of the old members, despite the fact that the votes of the old members, individually and collectively, constitute a smaller proportion of the total number of votes in the enlarged body.

In the empirical case studied, Luxembourg, apparently, was not aware of its dummy status on the EC Council of Ministers from 1958 to

1973. Nor, does it seem, was it aware that its voting power shot up not only in 1973 but again in 1981, when the council was expanded further; in 1986, however, its voting power will be cut by more than half. Astonishingly, from 1981 to 1985 Luxembourg attained the same status as Denmark and Ireland, countries with more votes and much larger populations.

The institutional designers of the council, I believe, were not acting irrationally but were simply ignorant. Believing that "the existing balance of power between the Member States would thus be broadly preserved" with the admission of Greece in 1981,[16] they succeeded—undoubtedly with good intentions—in jeopardizing it.

My suggestion is that, next time, careful formal analysis of voting power, which the EC has apparently eschewed, be carried out *prior* to implementing such changes. The fact that the bizarre effects of the 1973, 1981, and 1986 rule changes seem to have gone undetected underscores the capricious nature of constitution writing—done mostly by lawyers uninformed as to the significance of weights and decisions rules they set down—even today.

NOTES

1. Robin Farquharson, *Theory of Voting* (New Haven, Conn.: Yale University Press, 1969). Farquharson was the first to define and analyze sophisticated voting, which has been developed further in, among other places, Steven J. Brams, *Game Theory and Politics* (New York: Free Press, 1975), chap. 2; and Hervé Moulin, *The Strategy of Social Choice* (Amsterdam: North-Holland, 1983).
2. For a breakdown, see Steven J. Brams, *Paradoxes in Politics: An Introduction to the Nonobvious in Political Science* (New York: Free Press, 1976), pp. 168-175, which is based on Steven J. Brams and Frank C. Zagare, "Deception in Simple Voting Games," *Social Science Research* 6 (September 1977): 257-272.
3. For a more systematic analysis of deception voting strategies, see Brams and Zagare, "Deception in Simple Voting Systems"; and Brams and Zagare, "Double Deception: Two against One in Three-Person Games," *Theory and Decision* 13 (March 1981): 81-90.
4. For a simple demonstration in the three-candidate case, see Steven J. Brams, *The Presidential Election Game* (New Haven, Conn.: Yale University Press, 1978), pp. 199-202.
5. See Steven J. Brams, Dan S. Felsenthal, and Zeev Maoz, "New Chairman Paradoxes" (mimeographed, 1985); and Steven J. Brams, Dan S. Felsenthal,

and Zeev Moaz, "Chairman Paradoxes under Approval Voting" (mimeographed, 1985).

6. As used here, the idea of "control over outcomes" excludes the preferences of actors, which are central in the implicit formulation of power in the previous sections. Along with Nagel, I believe that a complete definition of power must include both preferences and outcomes (that is, power is the causation of outcomes by preferences), which I reintroduce in the game-theoretic analysis of Section 6.2. See Jack H. Nagel, *The Descriptive Analysis of Power* (New Haven, Conn.: Yale University Press, 1975), esp. chap. 3. Nagel links preferences and outcomes descriptively (that is, empirically), but not formally.
7. For a good collection of readings on the concept of power, see *Political Power: A Reader in Theory and Research.* ed. Roderick Bell, David V. Edwards, and R. Harrison Wagner (New York: Free Press, 1969).
8. John F. Banzhaf III, "Weighted Voting Doesn't Work: A Mathematical Analysis," *Rutgers Law Review* 19 (Winter 1965): 329-330, n. 31.
9. Ibid. For a review and comparison of different indices, and of their underlying assumptions about power, see Brams, *Game Theory and Politics,* chap. 5.
10. See Brams, *Game Theory and Politics,* pp. 176-182, for illustrations of the other voting power paradoxes.
11. This section is drawn in part from Steven J. Brams and Paul J. Affuso, "New Paradoxes of Voting Power on the EC Council of Ministers," *Electoral Studies* 4 (August 1985): 187-191.
12. For the original council comprising six members, it is not difficult to calculate the Banzhaf power values by hand, as illustrated in Sections 5.4 and 5.5. But for the subsequent larger councils, I would recommend writing a computer program to facilitate the calculations.
13. Steven J. Brams and Paul J. Affuso, "Power and Size: A New Paradox," *Theory and Decision* 7 (February/May 1976): 29-56.
14. L. S. Shapley and Martin Shubik, "A Method of Evaluating the Distribution of Power in a Committee System," *American Political Science Review* 48 (September 1954): 787-792.
15. These voting weights for Portugal and Spain, but a different decision rule (51 out of 76), were originally proposed in "Enlargement of the Community: Transitional Period and Institutional Implications," *Bulletin of the European Communities* 11 (Supplement, February 1978): 10-11.
16. Ibid., p. 10.

6 Threats and Deterrence

6.1 Introduction

Deterrence involves an exercise of power that may never be revealed or detected. By threatening untoward action against an opponent who initiates conflict, even at great potential cost to oneself, one seeks to deter the opponent from committing aggression in the first place. If the threat is successful, its deterring effect may be difficult to establish, because the opponent might have complied with it even if it had not been made.

Deterrence is the cornerstone of the national security policies of the superpowers and of other nations as well. The controversy over its viability has largely concerned the rationality of adhering to a policy that can lead to enormous destruction—perhaps even mutual annihilation—if the policy fails. In the case of the superpowers, the attacked party would seem foolhardy to bring upon itself a disastrous outcome if by compromising or—heaven forbid!—capitulating it could do better. By fighting to the bitter end, the attacked party would seem to violate the very canons of rationality on which deterrence rests. On the other hand, by caving in upon being threatened, or by indicating that it might cave in, it would seem to invite attack.

Hessel and I suggested that one possible resolution of the apparently conflicting conceptions of rationality embodied in deterrence is that carrying out a threat when deterrence fails, though irrational in the single play of a game, may well be rational in repeated play.[1] The reason is that a carried-out threat enhances one's credibility—because one has done the apparently irrational thing in a single play—so that over the long run one can develop a sufficiently fearsome reputation to deter future opponents. Thereby, while losing in the short run, one can win over time.

The use of threats as a deterrent illustrates aspects of power different from the paradoxical consequences of voting power discussed in the last chapter. For one thing, threats involve an attempt to influence others before taking action. For another, an ability to threaten may enable players to upset outcomes in certain games that otherwise would be stable, as will be shown.

In this chapter the telling effect that threats, and the ability to carry them out, may have on game outcomes will first be analyzed in a particular conflict—that between the Polish Communist party and Solidarity in the fateful 1980-1981 period. Solidarity was successful in asserting its influence early, but then the party recouped its dominant position. Each protagonist also sought to deter future challenges to its position by seizing the upper hand, even when this proved temporarily costly. Indeed, the Polish conflict illustrates the difference between "compellent" and "deterrent" threats, both of which involve one player's threatening to choose an inferior outcome (for both players) unless the other player accedes to the threat.

The question of making seemingly irrational threats comes up again in the analysis of the infamous Cuban missile crisis of October 1962. This crisis generated much tension between the superpowers over 13 days fraught with high drama and great danger. How the United States and the Soviet Union managed this crisis has been described in great detail, but such scrutiny has involved little in the way of formal and systematic strategic analysis. Three alternative perspectives will be presented on this crisis to demonstrate that it is open to different interpretations and that there is often no single "correct" view of a situation.

I hope not only to use game theory to shed light on the Cuban missile crisis but also to indicate how classical game theory can be revised so that it can better model the flow of moves by players over time. The founders of game theory recognized that their theory was a static one.[2] I shall attempt to show how assumptions of this theory can be reformulated in order better to capture the sequential calculations of players as they look ahead and try to anticipate consequences of their actions.

6.2 The Use of Threat Power in Poland, 1980-1981

The strategic situation in Poland in 1980-1981 was rife with threats and counterthreats.[3] In general terms, this situation may be viewed as a conflict pitting society against the state. More specifically, the conflict will be modeled as a game played between the leadership of the independent trade union, Solidarity, and the leadership of the Polish

Communist party/government/state (assumed to be the same for present purposes).

Although the threat of Soviet intervention was certainly a factor in the Polish game, I will not introduce the Soviet Union as a separate player. It is not clear to what extent Polish Communist party preferences in fact differed from Soviet ones, that is, to what extent these preferences would be changed if the Soviet influence—whether real or perceived—were absent. If it is true that Soviet preferences essentially paralleled those of the Polish Communist party, then the Soviets are best modeled not as a separate player but instead as a force on the side of the party that affected the balance of power in the game and hence the eventual outcome.

Each of the two sets of leaders may be treated as if it were a single decision maker. Of course, internal divisions within Solidarity and the party led to certain intraorganizational games; however, these subgames generally concerned not strategic choices on broad policy issues but rather tactical choices on more narrow operational questions. Focusing on the main game has the advantage of highlighting the most significant political-military choices each side considered, the relationship of these choices to outcomes in the game, and the dependence of these outcomes on threats and threat power.

The two players faced the following choices:

1. *Party.* Reject or accept the limited autonomy of plural social forces set loose by Solidarity. Rejection would, if successful, restore the monolithic structure underlying social organizations and interests; acceptance would allow political institutions other than the party to participate in some meaningful way in the formulation and execution of public policy.
2. *Solidarity.* Reject or accept the monolithic structure of the state. Rejection would put pressure on the government to limit severely the extent of the state's authority in political matters; acceptance would significantly proscribe the activities of independent institutions, and Solidarity in particular, to narrower nonpolitical matters, with only minor oversight over certain state activities.

Designate these strategies of both sides as "rejection" and "acceptance," and label them R and A, respectively. These strategies might also be designated "confrontation" and "compromise," but I prefer the former, more neutral labels because the disagreements were generally over specific proposals rather than general postures that the two sides struck.

The two strategies available to each side give rise to four possible outcomes:

1. *A-A.* Compromise that allows plural institutions but restricts their activities to nonpolitical matters, with negotiations leading to some sharing of political power being undertaken
2. *R-R.* Possibly violent conflict involving the entire society, opening the door to outside (mainly Soviet) intervention
3. *A (Solidarity)-R (Party).* Status quo ante, with tight restrictions on all activities of Solidarity and its recognition of the supremacy of party/state interests
4. *R (Solidarity)-A (Party).* Authorization of independent political activity and corresponding gradual reduction of the party/state role to implementation of public policy decisions made collectively

Again, the concern here is with strategic choices in the most general sense. The four outcomes represent what may be considered to be the four major scenarios pertinent to the Polish situation. Each of these scenarios can accommodate differences in details, but these differences seem more tactical than strategic. The four outcomes and the rankings I assign them are shown in the outcome matrix of Figure 6.1.

The party leadership repeatedly emphasized the unacceptability of any solution that would constrain its political power, which implies that

Figure 6.1 Polish Game, 1980-1981

		Party	
		R	A
Solidarity	A	Status quo ante (2,4)[a]	Compromise, with restrictions on political activities of Solidarity (3,3)[b]
	R	Violent conflict, with the possibility of outside intervention (1,2)	Authorization of independent political activity, leading to gradual reduction of the party's role (4,1)

Key: (x,y) = (Solidarity, party)
4 = best; 3 = next best; 2 = next worst; 1 = worst
A = acceptance; R = rejection
[a] Party's compellent threat outcome.
[b] Solidarity's deterrent threat outcome.
Circled outcome is an equilibrium.

its two worst outcomes are those associated with Solidarity's choice of R. Commenting on the Polish events, Bialer wrote, "Some [Party] leaders are more conservative and some more reformist; none, to our knowledge, questions the need to preserve the Party's monopoly."[4] The Eleventh Plenary Meeting of the Central Committee of the Polish United Workers Party (PUWP) was explicit on this point: "The Central Committee of PUWP unequivocally rejects . . . concepts of abandoning the leading role of the Party, of reducing this role to the ideological sphere and dispossessing the Party of the instruments of political power. This is the main danger."[5]

In fact, the available evidence indicates that the party preferred an all-out confrontation (R-R) to relinquishing its supremacy [R (Solidarity)-A (Party)]. Speaking at the Ninth Congress of PUWP, Deputy Prime Minister Mieczyslaw Rakowski announced: "To the enemies of socialism we can offer nothing but a fight, and not merely verbal at that." A later declaration of the Politburo reiterated that challenge: "We shall defend socialism as one defends Poland's independence. In this defense the state shall use all the means it deems necessary."[6] Finally, between its two best outcomes associated with Solidarity's choice of A, the party clearly preferred the status quo [A (Solidarity)-R (Party)] to compromise (A-A).

As for Solidarity, there is considerable evidence that it preferred the party's capitulation [R (Solidarity)-A (Party)] most, and violent conflict (R-R) least. In between, it preferred a compromise solution (A-A) to its own capitulation [A (Solidarity)-R (Party)]. Solidarity statements echoed this sentiment. Its chairman, Lech Walesa, said, "We don't want to change the socialist ownership of the means of production, but we want to be real masters of the factories. We were promised that many times before." Jacek Kuron, one of Solidarity's advisers, further clarified where the line on party activities should be drawn: "The party's leading role means the monopoly of power over the police forces, the army and foreign policy. All other matters must be open to negotiations with society."[7] In short, Solidarity preferred not to try to rob the party of its most significant functions, hoping to gain the party's acquiescence and thereby at least Solidarity's next-best outcome (A-A).

The reason for Solidarity's preference is evident. Solidarity was aware of the unacceptability of its best outcome [R (Solidarity)-A (Party)] to its opponent: "From the start, the Polish workers understood [that] to think of overthrowing the Party in Poland was madness, for it would inevitably lead to a Soviet invasion and the destruction of all liberties gained in the past ten or even twenty-five years." Addressing Solidarity members, Lech Walesa said: "Our country needs internal peace. I call on you to be prudent and reasonable." On another occasion he said, "[There are] fighters who want to fight at every opportunity; but we must understand that both the society and the union have had enough of confrontation . . . we ought not to go to the brink." Kuron concurred: "The goals of the government and of the democratic movement are completely opposite. But the struggle between the two tendencies, the totalitarian and the democratic one, are to be fought exclusively by peaceful means." [8] Thus, Solidarity preferred R only if the party chose A; if the party chose R, it preferred A.

The outcome matrix in Figure 6.1 reveals that the party has a dominant strategy of R, better for it whatever Solidarity chooses [(2,4) is better than (3,3) if Solidarity chooses A, (1,2) is better than (4,1) if Solidarity chooses R]. Anticipating this strategy choice of the party, Solidarity would prefer A to R, leading to (2,4), the unique equilibrium in this game; from the equilibrium outcome the unilateral deviation of either player would lead to a worse outcome. Because this equilibrium is the best outcome for the party, and only next worst for Solidarity, the game would appear to be inherently unfair to Solidarity.

Solidarity, however, can undermine this equilibrium if it possesses threat power, which may be either "deterrent" or "compellent," as will be described in the next paragraph.[9] In the model to be illustrated, such power enables a player to induce an outcome favorable to itself by threatening to implement a worse outcome for the other player; that this outcome is also worse for itself is why the threatener is assumed to possess greater power than the player that is threatened.

Solidarity can induce (3,3) by a *deterrent* threat: choosing A, it can threaten to choose R, and thus one of the party's two worst outcomes [(1,2) and (4,1)], unless the party accepts Solidarity's preferred outcome, (3,3). [Solidarity's best outcome, (4,1), is ruled out as a threat outcome because the party need never accept its worst outcome; it can always move to (1,2).] By contrast, the party can induce (2,4) with a *compellent* threat: choosing R, and refusing to move from it, it can force Solidarity to choose A over R, because Solidarity would prefer (2,4) to (1,2) if it must accede to one or the other.

Thus, whether Solidarity forces the party to accept (3,3) by threatening it with (1,2) or (4,1), or the party forces Solidarity to accept (2,4) by threatening it with (1,2), the outcome of the game turns on which player (if either) holds the balance of power. If Solidarity is the more powerful of the two players, or is at least perceived as such, it can implement (3,3). Otherwise, (2,4), as the equilibrium, would presumably obtain and would be reinforced should the party possess threat power. Note that Solidarity can implement its threat outcome, (3,3), by choosing its acceptance strategy A, relegating the choice of its rejection strategy R to a failure by the party to comply with its (deterrent) threat. This, of course, is the proverbial "speak softly and carry a big stick" policy, with the big stick invoked only if necessary.

The party, in contrast to Solidarity, can always do at least as well as Solidarity [(3,3), if it does not have threat power], and sometimes better [(2,4), if it does, or if neither player does], at least in terms of the comparative rankings of outcomes by the two players. Thus, a power asymmetry unfavorable to itself is not as serious for the party as for Solidarity, based on comparative rankings of the two threat outcomes. Moreover, because the party's threat is compellent, it can implement its best outcome simply by choosing and then maintaining its rejection strategy R, whereas Solidarity must first take a soft position (A) and then threaten escalation to its hard position (R), putting the onus for breakdown and subsequent disruption on itself.

This game-theoretic analysis based on threat power offers meaningful insights into the actual unfolding of events in Poland in 1980-1981. It seems that the party, perhaps stunned by the quick pace of developments and the widespread support for Solidarity after the August 1980 Lenin shipyard strike in Gdansk, did in fact consider Solidarity to be more powerful during the last part of 1980 and into the beginning of 1981. Reluctantly, it followed its acceptance strategy A; Solidarity, for its part, repeatedly emphasized the nonpolitical character of its demands (A) while threatening R, for the union's very existence was "based on adversary relations, not on a partnership." [10] As the economic situation worsened, however, the instability of the (3,3) compromise outcome became evident [recall that the stability of (3,3) rests on a power asymmetry favoring Solidarity], setting the stage for a test of strength between Solidarity and the party.

In March 1981, for the first time, the government used force against Solidarity. Although the force was limited in scope, its use can be interpreted as an attempt to switch to the party's rejection strategy R. Yet Solidarity chose to avoid confrontation, and the game remained at (3,3).

Although "the game was to leave the authorities with a semblance of power but to take its substance away," the events of March 1981 began to split Solidarity, strengthening proponents of the rejection strategy.[11] But the moderate leadership of Solidarity, and Walesa in particular, kept pointing to society's unwillingness to support the rejection strategy.[12] In doing so, the leadership cast doubt upon the viability of Solidarity's breakdown strategy, thereby undermining the union's power.

In December 1981 the party, apparently believing that the balance of power had shifted, switched decisively to its rejection strategy R, moving the game to *its* threat outcome, (2,4), by imposing martial law and jailing many Solidarity leaders in a massive crackdown. The stability of this outcome so far seems to validate the party's assessment of the balance of power favoring it in Poland—or at least demonstrates that Solidarity's power was not greater than its own. The fact that Solidarity has since shown no appetite to switch to R itself sustains the proposition that it remains reluctant to provoke violent conflict—especially one that it would almost surely lose—though the status quo ante seems less and less desirable for the party and may, perhaps, no longer be its best outcome.

6.3 The Cuban Missile Crisis as a Game of Chicken

There was never the threat of nuclear war in the Polish crisis, but that threat has been at least implicit in conflicts involving the nuclear powers (now six with a confirmed capability). Probably the most dangerous confrontation of this kind ever to occur was that between the United States and the Soviet Union in October 1962.[13] This confrontation, in what has come to be known as the Cuban missile crisis, was precipitated by a Soviet attempt to install in Cuba medium-range and intermediate-range nuclear-armed ballistic missiles capable of hitting a large portion of the United States.

After the presence of such missiles was confirmed on October 14, the CIA estimated that they would be operational in about ten days. A special executive committee of high-level officials was convened to decide on a U.S. course of action, and it met in secret for six days. Several alternatives were considered; these were eventually narrowed to the two discussed below.

The most common conception of this crisis is that the two superpowers were on a collision course. The game of chicken, which derives its name from a "sport" in which two drivers race toward each other on a narrow road, would appear to be an appropriate model of this conflict.

Under this interpretation, each player has the choice between swerving to avoid a head-on collision and continuing on the collison course. As applied to the Cuban missile crisis, with the United States and the Soviet Union the two players, the analogous courses of action and a ranking of the players' outcomes in terms of the game of chicken are shown in Figure 6.2.[14]

The goal of the United States was immediate removal of the Soviet missiles, and U.S. policy makers seriously considered two strategies to achieve this end:

1. *B.* A naval blockade, or "quarantine" as it was euphemistically called, to prevent shipment of further missiles, possibly followed by stronger action to induce the Soviet Union to withdraw those already installed

2. *A*. A "surgical" air strike to wipe out, insofar as possible, the missiles already installed, perhaps followed by an invasion of the island

The strategies open to Soviet policy makers were:

1. *W*. Withdrawal of their missiles
2. *M*. Maintenance of their missiles

Needless to say, the strategy choices and probable outcomes as presented in Figure 6.2 provide only a skeletal picture of the crisis as it developed over a period of 13 days. Both sides considered more than the two alternative courses of action listed here, as well as several variations on each. The Soviets, for example, demanded withdrawal of American missiles from Turkey as a quid pro quo for withdrawal of their missiles from Cuba, a demand ignored by the United States.

The second limitation of this picture is that there is no way to verify that the outcomes given in Figure 6.2 were probable, or valued in a manner consistent with the game of chicken. For example, if the Soviet Union had viewed an air strike on their missiles as jeopardizing their vital national interests, the A-W outcome (at the intersection of strategies A and W) may well have ended in nuclear war between the two sides, giving it the same value as A-M. Still another simplification relates to the assumption that the players chose their actions simultaneously, when in fact a continuous exchange in both words and deeds occurred over those fateful days in October.

Nevertheless, most observers of this crisis believe the two superpowers were on a collision course, a belief reflected in the title *Collison*

Figure 6.2 The Cuban Missile Crisis as a Game of Chicken

		Soviet Union	
		Withdrawal (W)	Maintenance (M)
United States	Blockade (B)	Compromise (3,3)	Soviet victory, U.S. defeat ((2,4))
	Air strike (A)	U.S. victory, Soviet defeat ((4,2))	Nuclear war (1,1)

Key: (x,y) = (United States, Soviet Union)
4 = best; 3 = next best; 2 = next worst; 1 = worst
Circled outcomes are equilibria.

Course, a book recounting this nuclear confrontation.[15] Most observers also agree that neither side was eager to take any irreversible step, such as a driver in a game of chicken might do by defiantly ripping out his or her steering wheel in full view of his or her adversary, thereby foreclosing the alternative of swerving.

Although in one sense the United States won by getting the Soviet Union to withdraw its missiles, at the same time Premier Nikita Khrushchev extracted from President John F. Kennedy a promise not to invade Cuba, which seems to indicate that the eventual outcome was a compromise solution of sorts. Moreover, even though the Soviet Union responded to the blockade and did not make its choice of a strategy independently of the U.S. choice, the fact that the United States held out the possibility of escalating the conflict to at least an air strike would seem to indicate that the initial blockade decision was not considered final—that is, the United States considered its strategy choices still open after imposing the blockade.

As I illustrated in the game-tree representation of Vashti's disobedience in Figure 2.2, one could model the Cuban missile crisis as a sequential game, in which the United States chooses first and the Soviet Union responds. Then, if the United States chooses blockade initially, it still has recourse to escalate to air strike if the Soviet Union proves intransigent (this representation will be developed in Section 6.5).

But first consider the choices facing the two players in the Figure 6.2 game. Each player obtains its next-best outcome (3) by choosing its strategy associated with compromise—if the other player also does. But both have an incentive to defect from this outcome to obtain their best outcomes of 4 by choosing their noncooperative strategies (A for the United States, M for the Soviet Union) when the other player chooses its cooperative strategy (B for the United States, W for the Soviet Union). Yet if both choose their noncooperative strategies, they bring nuclear war upon themselves.

Obviously both players want to avoid the disastrous (1,1) outcome, but the compromise (3,3) outcome is not an equilibrium because each player has an incentive to defect from it, as was just pointed out. In addition, neither player has a dominant, or unconditionally best, strategy: cooperation is best if the other player does not cooperate, but noncooperation is best if the other player cooperates.

True, the resulting outcomes, (4,2) and (2,4), are equilibria, because defection by either from these outcomes would lead to worse outcomes—(3,3) if the player obtaining 4 departs, (1,1) if the player obtaining 2 departs. But these two equilibria are not a *common* solution in the sense that the United States would prefer (4,2), the Soviet Union (2,4).

The competitive nature of these equilibria gives each player an incentive to threaten not to cooperate, hoping the other will concede by choosing cooperation so that the threatener can obtain its preferred equilibrium (and best outcome). As a model not only of the Cuban missile crisis but of other confrontation situations as well, chicken has often been used to explain why the superpowers have been willing to flirt with nuclear war on occasion.

Ostensibly, chicken would seem to mirror the intimidation implicit in a policy of deterrence whereby the United States and the Soviet Union threaten each other with mutual assured destruction (MAD) should either side attack the other. Yet the crisis seems better viewed as a breakdown, albeit temporary, in the more or less stable deterrence relationship between the superpowers that had held from World War II until that point.

To be sure, the Soviets undoubtedly concluded that it was no longer in their interest to continue this relationship because they calculated that any reprisals for installing the missiles would not be too severe. Presumably they reckoned that the probability of nuclear war was not high, thereby making it rational for them to risk provoking the United States. Although this thinking may have been more or less correct, there are good reasons for believing that U.S. policy makers viewed the game not as chicken at all, at least as far as they ranked the possible outcomes. In Figure 6.3 an alternative representation of the Cuban missile crisis is presented, retaining the same strategies for both players as given in the chicken representation (Figure 6.2) but assuming a different ranking of outcomes by the United States.[16] These may be interpreted as follows:

Figure 6.3 Payoff Matrix of Alternative Representation of the Cuban Missile Crisis

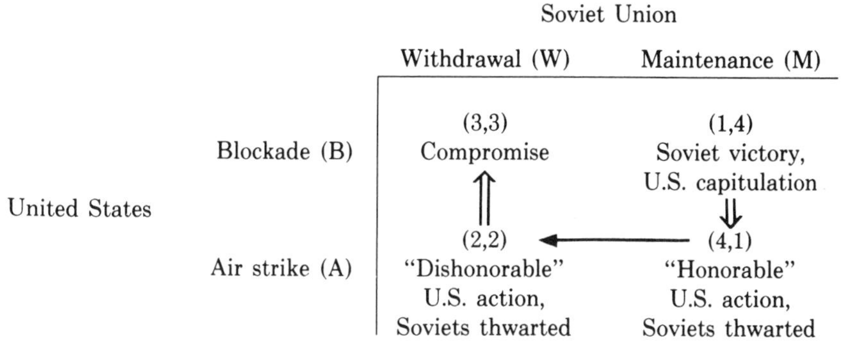

Key: (x,y) = (United States, Soviet Union)
4 = best; 3 = next best; 2 = next worst; 1 = worst
Arrows indicate rational moves of United States (vertical) and Soviet Union (horizontal); double arrows signify moving power of United States (see text).

1. *B-W*. The choice of blockade by the United States and withdrawal by the Soviet Union remains the compromise outcome for both players, (3,3).
2. *B-M*. In the face of a U.S. blockade, Soviet maintenance of their missiles leads to victory for the Soviet Union, its best outcome, and capitulation by the United States, its worst outcome, (1,4).
3. *A-W*. An air strike that destroys the missiles the Soviet Union was maintaining is an "honorable" action by the United States, its best outcome, and a defeat for the Soviet Union, its worst outcome, (4,1).
4. *A-M*. An air strike that destroys the missiles the Soviet Union was withdrawing is a "dishonorable" action by the United States, its next-worst outcome, and a defeat for the Soviet Union, its next-worst outcome, (2,2).

Even though an air strike thwarts the Soviet Union in the case of both outcomes A-W and A-M, I interpret outcome A-M to be a less damaging outcome for the Soviet Union because world opinion, it may be surmised, would condemn the air strike as an overreaction if there were clear evidence that the Soviet Union was in the process of withdrawing its missiles anyway. On the other hand, given no such evidence, a U.S. air strike, perhaps followed by an invasion, would probably be viewed by U.S. policy makers as a necessary, if not honorable, action to dislodge the Soviet missiles; by comparison, an air strike in the face of Soviet withdrawal would be unnecessary and, consequently, dishonorable. [If one viewed A-M not as the best outcome for the United States, as indicated in Figure 6.3, but instead as the next best—as (3,1), with the compromise outcome viewed as best, that is, as (4,3)—the subsequent analysis of deception possibilities in this game in Section 6.4 would not be affected.]

Before analyzing these possibilities, however, I shall offer a brief justification—mainly in the words of the participants—for the alternative ranking and interpretation of outcomes. The principal protagonists, of course, were Kennedy and Khrushchev, the leaders of the two countries. Their private communications over the 13 days of the crisis indicate that they both understood the dire consequences of precipitous action and shared, in general terms, an interest in preventing nuclear war. For the purposes of the present analysis, however, their specific preferences for each outcome are more relevant.

Given that the Soviet Union would withdraw its missiles, did the United States prefer an air strike (and possible invasion) to the blockade (followed by its eventual removal)? In responding to a letter from Khrushchev, Kennedy said: "If you would agree to remove these weapons from Cuba . . . we, on our part, would agree . . . (a) to remove promptly the quarantine measures now in effect and (b) to give assurances against an invasion of Cuba." [17] This statement is consistent with the alternative representation (Figure 6.3) of the crisis [since (3,3) is preferred to (2,2) by the United States] but not with the chicken representation (Figure 6.2) [since (4,2) is preferred to (3,3) by the United States].

Did the United States prefer an air strike to the blockade, given that the Soviet Union would maintain its missiles? According to Robert Kennedy, a close adviser to his brother during the crisis, "If they did not remove those bases, we would remove them." [18] This statement is consistent with the alternative representation [since (4,1) is preferred to (1,4) by the United States] but not consistent with the chicken representation [since (2,4) is preferred to (1,1) by the United States].

Finally, it is well known that several of President Kennedy's advisers felt very reluctant to initiate an attack against Cuba without exhausting less belligerent courses of action that might bring about removal of the missiles with less risk and greater sensitivity to American ideals and values. As Robert Kennedy put it, an immediate attack would be looked upon as "a Pearl Harbor in reverse, and it would blacken the name of the United States in the pages of history." [19] This statement is consistent with a U.S. ranking of the outcome A-W as next worst (2)—a dishonorable U.S. action in the alternative representation—rather than best (4)—a U.S. victory in the chicken representation.

If the alternative representation of the Cuban missile crisis is a more realistic representation of the participants' perceptions than the chicken representation, it still offers little in the way of explanation of how the compromise (3,3) outcome was achieved and rendered stable. After all, as in chicken, this outcome is not an equilibrium; but furthermore, in the alternative representation, unlike in chicken, no other outcome is stable either.

The instability of all outcomes in the Figure 6.3 game can be seen most easily by examining the cycle of preferences, indicated by the arrows between all pairs of adjacent outcomes (ignore for now the distinction between single and double arrows). These arrows show that at each outcome one player always has an incentive to move to another outcome—in the same row or column—because it can do better by such a move: the Soviet Union from (3,3) to (1,4); the United States from (1,4) to (4,1); the Soviet Union from (4,1) to (2,2); and the United States from (2,2) to (3,3).

Because one player always has an incentive to move from every outcome, none of the outcomes in the Figure 6.3 game is an equilibrium, as (4,2) and (2,4) are in Figure 6.2. Nor does either player have a dominant strategy; as in chicken, each player's best strategy depends on the strategy choice of the other player. Thus, for example, the United States prefers B if the Soviet Union chooses W, but A if it chooses M.

How, then, can one explain the choice of (3,3) in the Figure 6.3 game, given that this is a plausible reconstruction of the crisis? I shall suggest in Section 6.4 two qualitatively different sorts of explanation, one based on deception by the Soviet Union and the other based on the exercise of two different kinds of power by the United States. Then in Section 6.5 I shall present a game-tree reconstruction of sequential choices in the crisis.

6.4 Deception and Power in the Cuban Missile Crisis

By analogy to deception strategies in voting games (Section 5.3), define a player's *deception strategy* in a matrix game to be a false announcement of his or her preferences to induce the other player to choose a strategy favorable to the deceiver.[20] For deception to work, the deceived player must (1) not know the deceiver's true preference ranking (otherwise the deceiver's false announcement would not be believed), and (2) not have a dominant strategy (otherwise the deceived would always choose it, whatever the deceiver's announced preferences).

Given that conditions (1) and (2) are met, the deceiver, by announcing one of his or her two strategies to be dominant, can induce the deceived to believe he or she will always choose it. Anticipating this choice, the deceived will then be motivated to choose his or her strategy that leads to the better of the two outcomes associated with the deceiver's (presumed) dominant strategy.

As an illustration of the possible use of deception by the Soviet Union in the Figure 6.3 game, consider first the sequence of communications that led to a resolution of the crisis. As the crisis heightened, the Soviet Union indicated an increasing predisposition to withdraw rather than maintain their missiles if the United States would not attack Cuba and would pledge not to invade it in the future. Several statements by Premier Khrushchev support this interpretation of a shift in preferences. The first came in a letter to the British pacifist Bertrand Russell, the second in a letter to President Kennedy:

> If the way to the aggressive policy of the American Government is not blocked, the people of the United States and other nations will have to pay with millions of lives for this policy.
>
> If assurances were given that the President of the United States would not participate in an attack on Cuba and the blockade lifted, then the question of the removal or the destruction of the missile sites in Cuba would then be an entirely different question.[21]

Finally, in an almost complete about-face, Khrushchev, in a second letter to Kennedy, all but reversed his original position and agreed to remove the missiles from Cuba, though he demanded the quid pro quo alluded to earlier (which had been ignored by Kennedy):

> We agree to remove those weapons from Cuba which you regard as offensive weapons. . . . The United States, on its part, bearing in mind the anxiety and concern of the Soviet state, will evacuate its analogous weapons from Turkey.[22]

Khrushchev, who had previously warned (in his first letter to Kennedy) that "if people do not show wisdom, then in the final analysis they will come to clash, like blind moles," seemed, over the course of the crisis, quite ready to soften his original position.[23] This is not to say that his later statements misrepresented his true preferences—on the contrary, his language evoking the fear of nuclear war has the ring of truth to it. Whether he actually changed his preferences or simply retreated strategically from his earlier pronouncements, there was a perceptible shift from a noncooperative position (maintain the missiles regardless) to a cooperative position (withdraw the missiles if the United States cooperates).

Perhaps the most plausible explanation for Khrushchev's modification of his position is that there was, in Howard's terminology, a "deterioration" in his original preferences in the face of their possible apocalyptic game-theoretic consequences.[24] By interchanging, in effect, "3" and "4" in the Soviet preferences as shown in Figure 6.3, Khrushchev made W appear dominant, thereby inducing Kennedy also to cooperate (choose B). The resulting (3,3) outcome is next best for both players.

Whether Khrushchev deceived Kennedy or whether he actually changed his preferences, the effect is the same in inducing the compromise that was actually selected by both sides. Although there seems to be no evidence that conclusively establishes whether Khrushchev's shift was honest or deceptive, this question is not crucial to the analysis. True, I have developed the analysis in terms of rational deception strategies, but it could as well be interpreted in terms of genuine changes in preferences, given that preferences are not considered immutable.

Could the United States have deceived the Soviet Union to induce (3,3)? The answer is no: if the United States had made B appear dominant, the Soviet Union would have chosen M, resulting in (1,4); if the United States had made A appear dominant, the Soviet Union would have chosen W, resulting in (2,2). Paradoxically, because the United States could not through deception ensure an outcome better than its next worse (2)—whatever preferences it announced—it was in the interest of the United States to be deceived (or at least induced) so that (3,3) could be implemented.

More generally, in 5 of the 78 strictly ordinal 2 x 2 games, at least one player can do better as the deceived than as deceiver, so he or she profits by not knowing the preferences of the other player.[25] For this set of games the curious notion that "ignorance is strength"—or "ignorance is bliss"—seems well founded.

Is there any way that the United States, on its own, could have engineered the (3,3) outcome? Consider its choice of playing it safe by selecting its security-level strategy, which in this case is that associated with its next-worst outcome of 2. (A player's *security level* is the best outcome or payoff he or she can ensure, whatever strategy the other player chooses.) The choice of such a strategy to avoid its worst outcome (1) means selecting A; but if the Soviet Union also chooses its security-level strategy (W), the resulting outcome is (2,2), which is *Pareto-inferior,* or worse for both players, than (3,3).

If it is reasonable to assume that because the conflict occurred in the Caribbean—in the U.S. sphere of influence—the United States could exercise greater power than the Soviet Union, then there are means by which the United States could induce (3,3). Indeed, three kinds of power defined for 2 x 2 games—moving, staying, and threat—can all be used to implement (3,3).[26] In Section 6.2 I illustrated the use of threat power. I shall now describe and illustrate moving power; staying power will be discussed in Section 7.6.

Moving power is the ability of a player to continue moving in a game, like that in Figure 6.3, when the other player must eventually stop.[27] Assume that the United States has moving power, which I indicate by the vertical double arrows in Figure 6.3. This means that the United States will be able to hold out longer than the Soviet Union in the move-countermove cycle shown in Figure 6.3.

Eventually, then, the Soviet Union must break the cycle when it has the next move—at either (3,3) or (4,1), from which the single arrows emanate. Because the Soviet Union prefers (3,3), this is the moving power outcome the United States can eventually implement.

The threat power that the United States has in the Figure 6.3 game is of the deterrent variety:[28] by threatening to choose A, which includes the Soviet Union's two worst outcomes, the United States can induce the Soviet Union to choose W when the United States chooses B, resulting in (3,3). Even though the Soviet Union has an incentive to move from (3,3) to (1,4), as indicated by the top horizontal arrow, it would be deterred from doing so by the threat that if it did, the United States would choose its strategy A and stay there, inflicting upon the Soviet Union an outcome inferior to (3,3)—presumably (2,2), its better outcome in this row. Because it is rational for the Soviet Union to accede to this threat given that the United States has threat power, the possession of such power by the United States can also be used to implement (3,3).

It turns out that if the Soviet Union had moving or threat power in the Figure 6.3 game, it, too, could implement (3,3). Similarly, the possession of staying power by either player would also lead to the implementation of (3,3).[29]

Because whoever possesses any of the three kinds of power has no effect on the outcome that would be implemented in the Figure 6.3 game—(3,3)—such power is said to be *ineffective.* However, though ineffective, the impact of the power certainly is salutary in allowing the players to avoid a worse outcome—such as (2,2)—in a game in which neither player has a dominant strategy, there are no equilibria, and, consequently, there is no unequivocally rational choice in the absence of coercion.

Thus, as with being deceived, being influenced into choosing an outcome in a thoroughly unstable game, such as that in Figure 6.3, may not be unrewarding. Quite the contrary: one player's ability to deceive, or induce, the other player may be critical in effecting a compromise outcome.

6.5 A Sequential View of the Cuban Missile Crisis

Whether any of these forces was instrumental in resolving the Cuban missile crisis is hard to say with certitude. In fact, a much simpler calculation might have been made, based on the game tree (Section 2.2) and payoff matrix in Figure 6.4. Here I have altered the representation of the crisis once again to illustrate another plausible rendering of the alternatives and the preferences of the players for them.[30]

Starting at the top of the game tree, which is also called a *game in extensive form,* the United States can choose between an initial blockade and an immediate air strike (the later air-strike option is considered only if the Soviets do not cooperate by maintaining their missiles). If the United States chooses blockade, the Soviet Union can choose, as before, between maintaining and withdrawing its missiles. Finally, if by maintaining its missiles the Soviet Union chooses not to cooperate, then the United States can choose either no escalation—by continuing the blockade (blockade subsequently)—or escalation (to later air strike), assuming it had previously demurred.

For the preferences shown, which I shall not try to justify here because they are similar to those previously given for the players, start the backward-induction process (Section 2.2) at the bottom of the game tree. The United States would prefer—should play reach this point—air strike to blockade, so cut the latter branch to indicate that it would not be chosen and that (3,1) would be the outcome at this point.

Comparing at the next-higher level (4,3) with (3,1), which the United States would choose if it were given the choice at the bottom, the Soviet

Figure 6.4 Game Tree and Payoff Matrix of Sequential Choices in the Cuban Missile Crisis

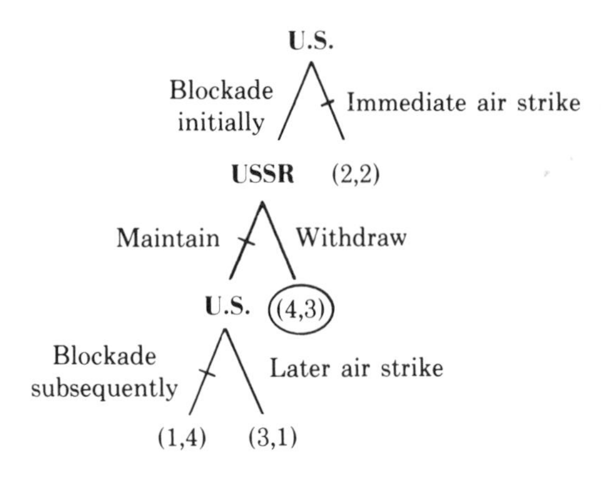

		USSR	
		Withdraw	Maintain
U.S.	Immediate air strike	(2,2)	(2,2)
	Blockade subsequently (if USSR maintains)	(4,3)	(1,4)
	Later air strike (if USSR maintains)	((4,3))	(3,1) ←Dominant strategy

Key: (x,y) = (United States, Soviet Union)
4 = best; 3 = next best; 2 = next worst; 1 = worst
Circled outcome in game tree is rational outcome (following path of uncut branches, starting at top); circled matrix outcome is an equilibrium.

Union would prefer (4,3), so cut the "maintain" branch. Finally, at the top of the tree, comparing (4,3)—which would move up based on the preceding analysis—with (2,2), the United States would prefer (4,3).

To ascertain rational choices of the players, reverse the backward-induction process. Starting at the top of the tree, the players would follow the uncut branches: the United States would blockade initially, and the Soviet Union would subsequently choose to withdraw. This, of course, is what happened in the crisis.

The matrix representation of this game tree, which is also called a *game in normal form,* is shown at the bottom of Figure 6.4. Note that the

United States has a dominant strategy, but the Soviet Union does not; anticipating the dominant choice of the United States—blockade initially, then air strike if the Soviet Union maintains (indicated only by "later air strike" in Figure 6.4)—the Soviet Union would prefer (4,3) to (3,1) and hence would choose to withdraw.

The resulting (4,3) outcome is the same rational choice as that deduced from the game-tree analysis (as it should be). The difference is that, in the normal or matrix form, strategies describe complete plans of action for all contingencies that may arise; by contrast, in the extensive or game-tree form, only single choices of the players are indicated at each move where the tree branches.

Whichever representation of the Cuban missile crisis one finds most congenial, together they illuminate different aspects of player choices in the crisis and problems that might have arisen in achieving a rational, stable outcome. The compromise reached—if that is what it was—has remained more or less intact for more than 20 years, which perhaps testifies to its durability.

Although the United States has had subsequent conflicts with Cuba, none has involved a confrontation with the Soviet Union on nearly the scale that occurred in 1962. (Indeed, at the height of the crisis President Kennedy estimated the chances of nuclear war to be between one-third and one-half.) [31] As a result of this crisis and the apprehension and fear it evoked, a "hot line" was established linking the heads of state in Washington, D.C., and Moscow, which has on occasion been used to try to prevent displays of brinkmanship from carrying the parties again so close to the nuclear precipice.

In the Figure 6.3 game, it is worth noting that the Soviet Union has a compellent threat (see Section 6.2 for another example) of choosing W and, by not moving from this strategy, inducing the United States to choose B, resulting in (3,3).[32] This is almost tantamount to its deception strategy, discussed in Section 6.4, of making W appear dominant and thereby inducing the United States to choose B.

Whether threats are explicit or implicit—as in the Figure 6.4 game tree, wherein it is assumed that the Soviet Union could anticipate a later air strike unless it responded to the blockade by withdrawing its missiles—they seem to have been part and parcel of the calculations of the protagonists in the Cuban missile crisis. Sorensen described this game-tree thinking when he reflected on American deliberations: "We discussed what the Soviet reaction would be to any possible move by the United States, what our reaction with them would have to be to that Soviet reaction, and so on, trying to follow each of those roads to their ultimate conclusion." [33]

In the next section I shall describe not only how one can specify this kind of thinking formally but also how one can ascertain whether there may be roads that lead to stable outcomes. Undoubtedly, what has been called for a generation the "delicate balance of terror" persists, at least to a degree; the intellectual challenge is now to find ways not to disturb it.[34]

6.6 Nonmyopic Equilibria and the Theory of Moves: The Search for Farsighted Solutions

I shall introduce in this section new rules of the game that allow one to distinguish short-term stable outcomes, or "myopic equilibria," from long-term stable outcomes, or "nonmyopic equilibria." By *myopic equilibria,* I mean those defined by Nash and described in Section 2.2 and subsequently in this book.[35] Nash's concept of equilibrium says, in effect, that a player considers only the *immediate* advantages and disadvantages of switching strategies. If neither player can gain immediately by a unilateral switch, the resulting outcome is stable, or a *Nash equilibrium.*

By contrast, Wittman and I, in defining a nonmyopic equilibrium, assume that a player, in deciding whether to depart from an outcome, considers not only the immediate effect of his or her actions but also the consequences of the other player's probable response, his or her own counterresponse, and so on—a sequence I shall refer to as *nonmyopic calculation.*[36] When neither player perceives a long-term advantage from departing from an initial outcome, this outcome is called a *nonmyopic equilibrium.* More specifically, Wittman and I assume that players look ahead and ascertain where they will end up if they depart from any outcome in an outcome matrix. They then compare the final outcome with the starting outcome; if they are better off at the starting outcome (taking account of their departures, possible responses to their departures, and so on), they will not depart in the first place. In this case, the starting outcome will be an equilibrium in an extended, or nonmyopic, sense.[37]

This new equilibrium concept presupposes a set of rules different from that assumed in classical game theory. Since a *game* is often defined to be the totality of the rules that describe it, the rules given in the next paragraph in fact define a new game.[38]

The *theory of moves* postulates four rules that define a *sequential game* (rule I is usually the only rule that the classical theory posits to govern the play of normal-form, or matrix, games):[39]

I. Both players simultaneously choose strategies, thereby defining an *initial outcome.*[40]

II. Once at an initial outcome, either player can unilaterally switch his or her strategy and change that outcome to a subsequent outcome in the row or column in which the initial outcome lies.
III. The second player can respond by unilaterally switching his or her strategy, thereby moving the game to a new outcome.
IV. The alternating responses continue until the player whose turn it is to move next chooses not to switch his or her strategy. When this happens, the game terminates, and the outcome reached is the *final outcome.*

Note that the sequence of moves and countermoves is *strictly alternating:* for example, first the row player moves, then the column player, and so on, until one stops, at which point the outcome reached is final.

How does a rational player determine whether he or she should move at a particular stage? I assume the player performs a backward-induction analysis, based on the game tree of possible moves that would be set off if he or she departed from the initial outcome. I shall develop this analysis in terms of players who make choices, according to the theory of moves, based on rules I-IV and other rules to be discussed later.

In the remainder of this section, the theory of moves will be illustrated with the game of chicken (Figure 6.2). I digress here from the analysis of the Cuban missile crisis; my purpose is not only to illustrate the theory of moves but also to show its effects in chicken, which is the game that seems best to model deterrence as a general phenomenon.[41] The theory's consequences are somewhat different in another infamous game, prisoners' dilemma, which I shall describe in Chapter 7.

Begin with two players, row and column. Assume that each of them initially chooses his or her compromise C strategy in chicken (Figure 6.2), resulting in the outcome (3,3). If row were to depart from this initial outcome and move the process to (4,2), column could then move it to (1,1); row could in turn respond by moving it to (2,4). These possible moves and the corresponding "stay" choices at each node are illustrated in Figure 6.5.

To determine rational choices for the players to make at each node of the game tree, starting at (3,3), it is necessary to work backward up the game tree in Figure 6.5, as shown in Section 6.5. Consider row's choice at (1,1). Because he or she prefers (3,4) to (1,1), I indicate that "stay" at (1,1) would not be chosen by cutting its branch, should the process reach the node at (1,1). Instead, outcome (2,4) would be chosen, which can be mentally substituted for the endpoint, row at (1,1).

Working backward again, compare (4,2) at the "stay" branch with (2,4) at the "move" branch (given the previous substitution). Since column would prefer (2,4), the "stay" branch at this node is cut, and (2,4)

Figure 6.5 Game Tree of Moves, Starting with Row, from (3,3) in Chicken

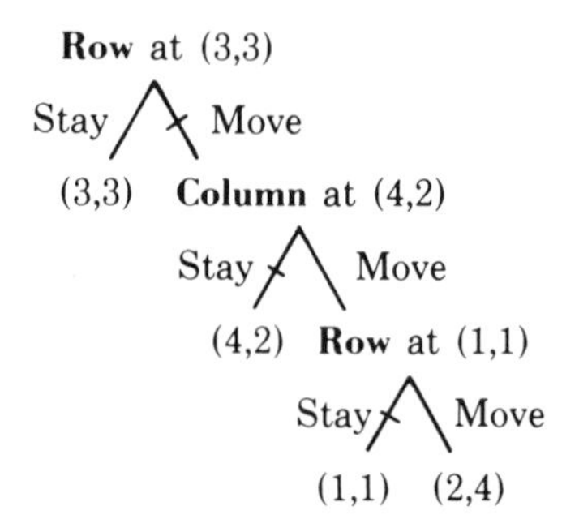

Key: (x,y) = (row, column)
4 = best; 3 = next best; 2 = next worst; 1 = worst
Circled outcome is rational (following path of uncut branches, starting at top).

moves up to a final comparison with (3,3) at the top node. At this node row would prefer (3,3), so the "move" branch at the top node is cut, and (3,3) is therefore the final outcome that survives the cuts.

In other words, there would be no incentive for row to depart from (3,3), anticipating the rational choices of players at subsequent nodes in the game tree of Figure 6.5. Similarly, because of the symmetry of the game, there would be no incentive for column to depart from (3,3). When the final outcome coincides with the initial outcome, as it does in the case of (3,3), it is a nonmyopic equilibrium.[42]

For the other three outcomes in chicken, there is no corresponding incentive for both players to stay, should any one be the initial outcome of the game. For example, a game-tree analysis starting at (1,1) reveals that row would have an incentive to depart to (2,4), and column to (4,2). After either departure, the process would terminate because the player with the next move—column at (2,4), row at (4,2)—would obtain his or her best outcome. But if (2,4) or (4,2) were the initial outcome, rational departures by row from (2,4) and column from (4,2) would carry the process to (1,1), whence it would go to (4,2) if row departed initially, (2,4) if column departed initially, and then stay for the reason just given.

But there is a complication starting at (1,1). Clearly, row would prefer that column depart first from (1,1), yielding (4,2), and column would prefer that row depart first, yielding (2,4). Since one cannot say a priori which player would be able to hold out longer at (1,1), forcing the other to move first, I designate the final outcome, starting at (1,1), as "(2,4)/(4,2)"—either is possible.

It is easy to show that if (2,4) were the initial outcome, the final outcome according to the game-tree analysis would be (4,2), and (2,4) if (4,2) were the initial outcome. This is because the player obtaining his or her next-worst outcome (2) can, by moving the process to outcome (1,1), force the other player to move to the outcome best for himself or herself [(4,2) for row, (2,4) for column]. In either case, the player obtaining his or her best outcome (4) at (2,4) or (4,2) would seem to have no incentive to depart to the inferior outcome, (3,3).

Yet an objection can be raised to this reasoning: the player who obtains 4 initially, knowing that he or she will be reduced to 2, has an incentive to move the process to (3,3) first, where the player obtains his or her next-best outcome rather than next-worst. Moreover, once at (3,3), the process would stop since a subsequent move by, say, row to (4,2) would move the process to (1,1) and thence to (2,4), which is inferior to (3,3) for row.

The countermove to (3,3) by the player whose best outcome is at (2,4) or (4,2) would appear to introduce a new kind of rational calculation into the analysis, namely, what the other player will do if one does not seize the initiative. True, I implicitly assumed earlier that each player, separately, would ascertain the final outcome only for himself or herself; yet it seems reasonable that each player would consider not only the consequences of his or her departing from an outcome but also the consequences of the other player's departing. Because each player could do better by holding out at, say, (1,1), each presumably would strive to delay his or her departure, hoping to force the other player to move first.

The situation starting at (2,4) or (4,2) is the reverse for the players. Although the game-tree analysis shows that, say, row should not move from (4,2) to (3,3), row's recognition of the fact that column can move the process to (2,4) would induce him or her to try to get the jump on column by moving first to (3,3). In contrast, at (1,1) each player has an incentive to hold out rather than scramble to leave this outcome first.

In either event, a rational choice is dictated not only by one player's own game-tree analysis but by that of the other player as well, which may cause a player to overide his or her own (one-sided) rational choice. Henceforth, I assume that a final outcome reflects the *two-sided* analysis that both players would make of each other's rational choices in addition to their own.

In the case of outcomes (2,4) and (4,2), it is impossible to say a priori which player would be successful in departing first. Accordingly, as in the case of (1,1), it seems best to indicate a *joint* final outcome of (4,2)/(3,3) starting from (2,4), and of (2,4)/(3,3) starting from (4,2).

In summary, the final outcomes of chicken, given that the players make rational choices according to a two-sided game-tree analysis and

Figure 6.6 Revised Chicken, with Final Outcomes

		Column	
		C	$\overline{C}$
Row	C	(3,3) (circled)	(4,2)/(3,3)
	$\overline{C}$	(2,4)/(3,3)	(2,4)/(4,2)

Key: (x,y) = (row, column)
4 = best; 3 = next best; 2 = next worst; 1 = worst
Circled outcome is a Nash equilibrium.

the four rules specified previously, are as follows for each initial outcome:

Initial Outcome	Final Outcome
(3,3)	(3,3)
(1,1)	(2,4)/(4,2)
(4,2)	(2,4)/(3,3)
(2,4)	(4,2)/(3,3)

If one substitutes the final outcomes for the initial outcomes in the payoff matrix of Figure 6.2, the new game shown in Figure 6.6 results. The outcomes of this game may be thought of as those that are obtained if the four rules of sequential play, coupled with rational choices based on a two-sided game-tree analysis, are operative.

In the preliminary analysis of this game, assume that each of the two outcomes in the joint pairs is equiprobable.[43] Then, in an expected-value sense, C dominates $\overline{C}$ for each player. If column, for example, chooses C, (3,3) is better for row than (2,4)/(3,3), which half the time will yield (2,4); if column chooses $\overline{C}$, (4,2)/(3,3) is better for row than (2,4)/(4,2) because, though the (4,2)'s effectively cancel each other out, (3,3) is preferred to (2,4) half the time.

Strictly speaking, however, for C to dominate $\overline{C}$ in *every play* of the game, it is necessary to make two assumptions: (1) whenever column chooses $\overline{C}$, if $\overline{C}$ for row yields (4,2) as a final outcome, so does C for row; (2) there is some possibility, however small, that the choice of $\overline{C}$-$\overline{C}$ by the players yields (2,4). In this manner, row's choice of C is always at least as good as, and sometimes better than, the choice of $\overline{C}$.

Assumption (1) above is the crucial assumption. It says, in effect, that whenever column chooses $\overline{C}$ in the original game of chicken, and row can hold out longer at (1,1) by choosing $\overline{C}$—forcing the final outcome to be (4,2)—row can preempt column at (2,4) by choosing $\overline{C}$, yielding the final outcome (4,2). In other words, if row is the "stronger" player at (1,1), he or she is also the "quicker" player at (2,4), because he or she is able to move the process to (1,1) before column moves it to (3,3).

The guarantee of dominance provided by assumptions (1) and (2) seems as reasonable as the expected-value assumption, which says that given the equiprobability of the two outcomes in the joint pairs, C dominates $\overline{C}$ on the average. Either way, rational players in chicken, anticipating the final outcomes shown in Figure 6.6, will both choose their dominant strategy C. Thereby the four rules of sequential play specified earlier induce the compromise (3,3) outcome in chicken, which is circled in Figure 6.6.

To be sure, in situations of nuclear deterrence, it seems implausible that players would consider actually moving through the (1,1) outcome, though they might threaten to do so. In situations of conventional deterrence, on the other hand, such punishing behavior might be considered appropriate to bolster one's reputation for toughness and thereby to enhance one's future credibility.

As I shall show in Chapter 7, the justification—according to the theory of moves—of the (3,3) outcome in a prisoners' dilemma model of traps relies on different comparisons. The logic underlying this theory, I believe, offers some hope for the resolution of crises by farsighted players; expected-payoff calculations may also play a role, as I shall demonstrate, if players have substantial detection capabilities.

6.7 Conclusions

To deter an adversary, a player must make the choice of aggression sufficiently unattractive, through the threat of retaliation if aggression should occur, that the adversary will refrain from aggression in the first place. In the Polish conflict of 1980-1981 between Solidarity and the Communist party, as I modeled it, there was a unique equilibrium outcome that was favorable to the party; yet Solidarity had the ability to threaten a worse outcome (for both players).

Solidarity's threat was in fact credible in the beginning, and the union temporarily gained some concessions from the party. Yet these were not to last, because the party, with Soviet support, could also use threat power to redress the imbalance and reinstitute the equilibrium, which it eventually did.

Threats are effective if they are backed up by the wherewithal to carry them out. But the threatened player must also recognize that the threatener will act—even to his or her own detriment—if the threat is ignored. Given both the power and the resolve to implement threats, should this prove necessary, threats can serve either to deter or to compel desired behavior, as the Polish conflict illustrated in the 1980-1981 period.

The problem with threats, of course, is that the threatener as well as the adversary may be hurt if the threatener is forced to carry out the threat, thereby making the threat appear quite incredible. This problem is no better illustrated than in the game of chicken, my initial model for the Cuban missile crisis.

The Soviet Union was perhaps banking on the incredibility of nuclear escalation, implicit in a policy of MAD, when it attempted to install missiles in Cuba. If deterrence failed initially, however, escalation short of nuclear war—or even an air strike—eventually succeeded.

I questioned whether the game played out in October 1962 between the superpowers was chicken, primarily to underscore the difficulty of identifying players' preferences in real-world games. The three different representations given of the Cuban missile crisis testify to the need to ponder strategic conflicts from different perspectives that take account as much of the perceptions of players as of their true preferences.

In the fashion of *Rashomon* (a Japanese movie that portrays four different versions of a rape), each perspective gives new insights. It is especially instructive to see how sensitive rational outcomes are to the different reconstructions on which each is based and to examine the relationship of these to the actual outcome.

A discrepancy between one's preferences and the perception of these by an opponent highlights another strategy players can adopt. Lack of complete information in a game may induce one player to try to deceive the other. There is evidence in the Cuban missile crisis that a deception strategy was tried and possibly succeeded in abetting a compromise.

Conflicts are dynamic events that unfold over time. I suggested that the theory of moves, and the nonmyopic equilibria based on this theory, may enable one to uncover stability in certain games that classical game theory hides or places elsewhere.

The importance of different kinds of power in games is illuminated by this theory. In political science and international relations, power has proved to be a very elusive concept, even at a theoretical level (Section 5.1), but definitions that tap different aspects of power—moving and threat in this chapter—come to the fore naturally in sequential games in which players can make rational moves and countermoves. Their jockeying to implement favorable outcomes epitomizes rational politics in games relatively free of constraints (unlike those played in voting bodies).

There may, nevertheless, be different constraints on *moves* in such games, and these can provide different measures of power. I have not attempted here to explore the full dimensions of the constraints that one player can impose upon another, having dealt with this topic elsewhere.[44] Rather, my purpose here has been to try to give certain strategic insights into deterrence, in part through the analysis of the use of threats and related maneuvers in two crises.

NOTES

1. Steven J. Brams and Marek P. Hessel, "Threat Power in Sequential Games," *International Studies Quarterly* 28 (March 1984): 23-44. See also Frank C. Zagare, "Toward a Reformulation of the Theory of Mutual Deterrence," *International Studies Quarterly* 29 (June 1985).
2. John von Neumann and Oskar Morgenstern, *Theory of Games and Economic Behavior,* 3d ed. (Princeton, N.J.: Princeton University Press, 1953), pp. 44-45.
3. This section draws on material in Brams and Hessel, "Threat Power in Sequential Games," which contains additional background information on the Polish crisis and also a formalization of the game-theoretic analysis of threats to be described later.
4. Severyn Bialer, "Poland and the Soviet Imperium," *Foreign Affairs* 59 (1981): 530.
5. *Nowy Dziennik,* June 17, 1981, p. 2; these and other translations of Polish language publications are by Marek P. Hessel.
6. *Nowy Dziennik,* July 21, 1981, p. 2, and September 22, 1981, p. 2.
7. *Time,* December 29, 1980, pp. 20, 29.
8. Neal Ascherson, *The Polish August: The Self-Limiting Revolution* (New York: Viking, 1982), pp. 18-19; *Time,* December 29, 1980, p. 2; *Solidarnosc,* April 10, 1981, p. 1; and interview with Jacek Kuron, *Telos* 57 (1981): 93-97.
9. The distinction between deterrence and compellence was originally suggested in Thomas C. Schelling, *Arms and Influence* (New Haven, Conn.: Yale University Press, 1966), but not formalized in this work.
10. Tadeusz Szafar, "Brinkmanship in Poland," *Problems of Communism* 30 (May/June 1981): 79.
11. Richard M. Watt, "Polish Possibilities," *New York Times Book Review,* April 25, 1982, p. 11.
12. Walesa was opposed to the strike because he considered the Polish situation too precarious, both internally and externally. Defending his position, he said, "We must move step by step, without endangering anybody. . . . I am not saying that there will be no confrontation over some important issue . . . but now the society is tired and wants no confrontation." *Solidarnosc,* April 10, 1981, p. 1.
13. This and the next section draw on material in Steven J. Brams, "Deception in 2 x 2 Games," *Journal of Peace Science* 2 (Spring 1977): 171-203.
14. Henceforth I shall assume in this crisis that the superpowers can be

considered unitary actors, though this is an obvious simplification. It is rectified in part by constructing other models that emphasize different features, as Allison has done. See Graham T. Allison, *Essence of Decision: Explaining the Cuban Missile Crisis* (Boston: Little, Brown, 1971).

15. Henry M. Pachter, *Collison Course: The Cuban Missile Crisis and Coexistence* (New York: Praeger, 1963). Other books on this crisis include Elie Abel, *The Missile Crisis* (Philadelphia: Lippincott, 1966); Allison, *Essence of Decision;* Robert F. Kennedy, *Thirteen Days: A Memoir of the Cuban Missile Crisis* (New York: W. W. Norton, 1969); Robert A. Divine, ed., *The Cuban Missile Crisis* (Chicago: Quadrangle, 1971); Abram Chayes, *The Cuban Missile Crisis: International Crises and the Role of Law* (New York: Oxford University Press, 1974); Herbert Dinerstein, *The Making of the Cuban Missile Crisis, October 1962* (Baltimore: Johns Hopkins University Press, 1976); and David Detzer, *The Brink: Story of the Cuban Missile Crisis* (New York: Crowell, 1979).
16. Still a different 2 x 2 game is proposed in Glenn H. Snyder and Paul Diesing, *Conflict among Nations: Bargaining, Decision Making, and Systems Structure in International Crises* (Princeton, N.J.: Princeton University Press, 1977), pp. 114-116; an "improved metagame analysis" of the crisis is presented in Niall M. Fraser and Keith W. Hipel, "Dynamic Modelling of the Cuban Missile Crisis," *Conflict Management and Peace Science* 6 (Spring 1982-1983): 1-18.
17. Allison, *Essence of Decision,* p. 228.
18. Kennedy, *Thirteen Days,* p. 170.
19. Theodore C. Sorensen, *Kennedy* (New York: Harper and Row, 1965), p. 684.
20. Brams, "Deception in 2 x 2 Games"; see also Steven J. Brams and Frank C. Zagare, "Deception in Simple Voting Games," *Social Science Research* 6 (Septcmber 1977): 257-272; Steven J. Brams and Frank C. Zagare, "Double Deception: Two against One in Three-Person Games," *Theory and Decision* 13 (1981): 81-90; Frank C. Zagare, "A Game-Theoretic Analysis of the Vietnam Negotiations: Preferences and Strategies 1968-1973," *Journal of Conflict Resolution* 21 (December 1977): 663-684; and Frank C. Zagare, "The Geneva Conference of 1954: A Case of Tacit Deception," *International Studies Quarterly* 23 (September 1979): 390-411. A useful compilation of material on deception, both in theory and practice, is given in *Strategic Military Deception,* ed. Donald C. Daniel and Katherine L. Herbig (New York: Pergamon, 1982).
21. Divine, *Cuban Missile Crisis,* pp. 38, 47.
22. Ibid., p. 49.
23. Ibid., p. 47.
24. Nigel Howard, *Paradoxes of Rationality: Theory of Metagames and Political Behavior* (Cambridge, Mass.: MIT Press, 1971), pp. 148, 199-201.
25. Brams, "Deception in 2 x 2 Games." The 78 2 x 2 games, in which each player can strictly rank the four outcomes from best to worst, are structurally distinct in the sense that no interchange of rows, columns, or players can transform one of these games into any other.

26. Steven J. Brams, *Superior Beings: If They Exist, How Would We Know? Game-Theoretic Implications of Omniscience, Omnipotence, Immortality, and Incomprehensibility* (New York: Springer-Verlag, 1983).
27. Steven J. Brams, "Omniscience and Omnipotence: How They May Help—or Hurt—in a Game," *Inquiry* 25 (June 1982): 217-231.
28. Brams and Hessel, "Threat Power in Sequential Games."
29. Steven J. Brams and Marek P. Hessel, "Staying Power in 2 x 2 Games," *Theory and Decision* 15 (September 1983): 279-302.
30. This game was suggested in a personal communication from Philip D. Straffin, Jr. (1976).
31. Sorensen, *Kennedy,* p. 705.
32. Brams and Hessel, "Threat Power in Sequential Games," pp. 35-36.
33. Quoted in Ole R. Holsti, Richard A. Brody, and Robert C. North, "Measuring Affect and Action in International Reaction Models: Empirical Materials from the 1962 Cuban Crisis," *Journal of Peace Research* 1 (1964): 188.
34. Albert Wohlstetter, "The Delicate Balance of Terror," *Foreign Affairs* 37 (January 1959): 209-234. However, Intriligator and Brito argue, on the basis of a dynamic model of a missile war from which they derive conditions for stable deterrence, that the chances of war outbreak have, paradoxically, been reduced because of the recent U.S.-Soviet quantitative arms race. See Michael D. Intriligator and Dagobert L. Brito, "Can Arms Races Lead to the Outbreak of War?" *Journal of Conflict Resolution* 28 (March 1984): 63-84.
35. John Nash, "Non-cooperative Games," *Annals of Mathematics* 54 (1951): 286-295.
36. Steven J. Brams and Donald Wittman, "Nonmyopic Equilibria in 2 x 2 Games," *Conflict Management and Peace Science* 6 (Fall 1981): 39-62. Applications of nonmyopic equilibria to two Middle East crises are made in Frank C. Zagare, "Nonmyopic Equilibria in the Middle East Crisis of 1967," *Conflict Management and Peace Science* 5 (Spring 1981): 139-162; and Frank C. Zagare, "A Game-Theoretic Evaluation of the Cease-Fire Alert Decision of 1973," *Journal of Peace Research* 20, no. 1 (1983): 73-86. In the latter article, Zagare also applies a related concept of game-theoretic stability that is analyzed in Steven J. Brams and Marek P. Hessel, "Absorbing Outcomes in 2 x 2 Games," *Behavioral Science* 27 (October 1982): 393-401. These applications are also discussed in Frank C. Zagare, *Game Theory: Concepts and Applications* (Beverly Hills, Calif.: Sage, 1984), chap. 1, and Steven J. Brams, *Superpower Games: Applying Game Theory to Superpower Conflict* (New Haven, Conn.: Yale University Press, 1985), chap. 2.
37. Brams and Wittman, "Nonmyopic Equilibria in 2 x 2 Games," pp. 42-43.
38. Brams, *Superior Beings,* p. 75.
39. For a full development of this theory, see ibid.
40. By *strategy* I mean here a course of action that can lead to any of the outcomes associated with it, depending on the strategy choice of the other player; the strategy choices of both players define an outcome at the intersection of their two strategies. While the subsequent moves and countermoves of players could also be incorporated into the definition of a strategy

meaning a complete plan of responses by a player to whatever choices the other player makes in a game (Section 2.2)—this would make the normal (matrix) form for the sequential game unduly complicated and difficult to analyze. Hence, I use "strategy" to mean the choices of players that lead to an initial outcome, and "moves" and "countermoves" to refer to their subsequent sequential choices, as allowed by rules II-IV.

41. Steven J. Brams, *Superpower Games: Applying Game Theory to Superpower Conflict* (New Haven, Conn.: Yale University Press, 1985), chap. 1.
42. Brams and Wittman, "Nonmyopic Equilibria in 2 x 2 Games"; see also D. Marc Kilgour, "Equilibria for Far-sighted Players," *Theory and Decison* 16 (March 1984): 135-157; D. Marc Kilgour, "Anticipation and Stability in Two-Person Non-Cooperative Games," in *Dynamic Models of International Conflict,* ed. Michael D. Ward and Urs Luterbacher (Boulder, Colo.: Lynne Rienner, 1985); and Frank C. Zagare, "Limited Move Equilibria in 2 x 2 Games," *Theory and Decision* 16 (January 1984): 1-19, for extensions and emendations in the concept of nonmyopic equilibrium. To ensure that a final outcome is reached, either at the start or before there is cycling back to the initial outcome, the definition of a nonmyopic equilibrium also includes a termination condition, or stopping rule. This condition specifies that if there exists a node in the game tree such that the player with the next move can ensure his or her best outcome by staying at it, he or she will. This condition is satisfied by the compromise (3,3) outcome in both chicken and prisoners' dilemma (discussed in Section 7.2).
43. The equiprobability assumption is not crucial; it is made to illustrate the calculation of expected values and is contrasted with other assumptions given in the next paragraph of the text.
44. Brams, *Superior Beings.*

7 Traps: No-Win Situations

7.1 Introduction

In a world ridden with conflict, one often hears the lament that our circumstances would be much improved and our lives would be happier if only people could learn to live with each other. Some who offer this lament believe that showing the unenlightened the benefits of cooperation would quickly convince them that they indeed have much to gain from friendlier relations and agreements that work to the benefit of all. To one leading student of international conflict, the heart of the problem is that people do not stop to think:

> Why are so many nations reluctantly but steadily increasing their armaments as if they were mechanically compelled to do so? Because . . . they follow their traditions . . . and instincts . . . and because they have not yet made a sufficiently strenuous intellectual and moral effort to control the situation.[1]

This world view is fundamentally optimistic, for it suggests that with the proper effort—however strenuous—the belligerent intentions of human beings can be defused and wars can be shown to be unproductive.

Other observers, like Lorenz, take a more pessimistic view of human nature and see in human actions an innate animal aggressiveness:

> Human behavior . . . far from being determined by reason and cultural tradition alone, is still subject to all the laws prevailing in all phylogenetically adapted instinctive behavior. . . . The evil effects of the human aggressive drives, explained by Sigmund Freud as the results of a special death wish, simply derive from the fact that in prehistoric times intra-

> specific selection bred into man a measure of aggression drive for which, in the social order of today, he finds no adequate outlet.[2]

The only hope Lorenz held out for the human race was in the displacement of its aggressive energy into socially desirable outlets such as sports contests and artistic or scientific endeavors.

A third school of thought, while viewing humans as innately aggressive and acquisitive, sees their salvation in the creation of institutions capable of enforcing a social contract. Seventeenth- and eighteenth-century political philosophers such as Hobbes and Rousseau believed that an internalized morality offered no escape from the anarchy and brutality of a world in which everybody pursued their self-interests. To rechannel the selfish and narrow interests of individuals toward more productive social ends would require a centralized authority capable of—in Rousseau's phrase—"forcing men to be free."

Similarly, Hobbes characterized the state as a "leviathan" because of the great power that would have to be vested in it to overcome formidable opposition by an amoral populace.[3] Rousseau recognized the problem of achieving social consensus. He distinguished the "will of all," reflecting the selfish interests of individuals, from the socially desirable "general will." To bring these distinct wills into line, he proposed a systematic program of "civil religion" to educate the populace on the virtues of submitting to the higher authority of the state.[4]

Although the classical political theorists clearly recognized the potential conflict between the interests of individuals and the interests of the community or state, they nevertheless were unclear about the specific conditions that would lead to the formation of the state. If regulated social life is better than a state of nature, must a populace be strong-armed into submission or indoctrinated through a program of civil religion? Or can society achieve beneficial ends without resorting to the use of force or some form of indoctrination?

Despite a mountain of empirical research on individual and social behavior, today's answers to these questions are still not very good. Perhaps better understood are the structural conditions that give rise to a divergence of individual and group interests. By focusing on these conditions, of course, one sidesteps the question of whether human beings are innately aggressive. But because this question seems unlikely to be resolved in the near future, it will be more profitable to analyze situations in which cooperation among two or more actors is difficult, if not impossible, to achieve.

Sadly, these situations occur despite the benefits the actors involved could achieve were they to act cooperatively. In this chapter I shall offer a simplified representation and give two extended examples of *traps,* or games in which the players' apparently rational strategies lead them to a collectively worse outcome than had they made other choices.[5] The first example is the superpower arms race, the second the White House tapes case that arose out of President Richard Nixon's refusal to release tapes of his conversations relating to the Watergate break-in and its aftermath. Conflicts that involve more than two players may also have a no-win character, and these will be briefly discussed at the end of this chapter.

7.2 Prisoners' Dilemma

The simplest example of a trap is the game of prisoners' dilemma, which is illustrated by the payoff matrix of Figure 7.1. The situation may be described by the following story, attributed to A. W. Tucker. Two persons suspected of being partners in a crime are arrested and placed in separate cells so that they cannot communicate with each other. Without a confession from *at least one* suspect, the district attorney does not have sufficient evidence to convict them of the crime. To try to extract a confession, the district attorney tells each suspect the following consequences of their (joint) actions:

1. If one suspect confesses and the partner does not, the one who confesses can go free (gets no sentence) for cooperating with the state, but the other gets a stiff 10-year sentence: (0,-10) and (-10,0) in the payoff matrix.
2. If both suspects confess, both get reduced sentences of 5 years: (-5,-5) in the payoff matrix.
3. If both suspects remain silent, both go to prison for 1 year on a lesser charge: (-1,-1) in the payoff matrix.

Figure 7.1 Prisoners' Dilemma

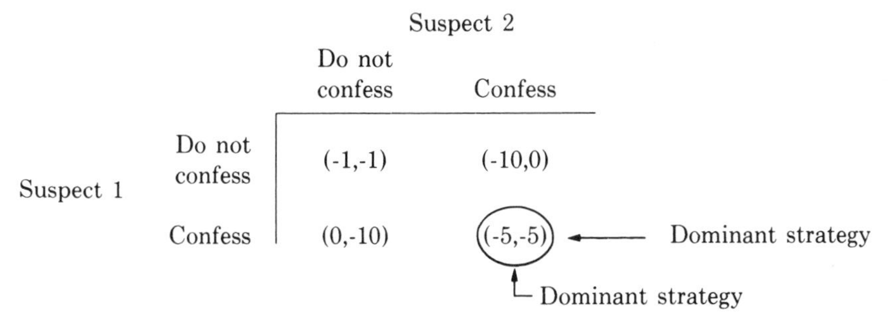

Key: (x,y) = (suspect 1, suspect 2)
Numbers in outcomes represent years in prison.
Circled outcome is Nash equilibrium.

What should each suspect do to save his or her own skin, assuming there is no compunction against squealing on a partner?[6] Observe first that if either suspect confesses, it is advantageous for the other suspect to do likewise to avoid his or her very worst outcome of 10 years in prison (-10). The rub is that the idea of confessing and receiving a moderate sentence of 5 years is not at all appealing, even though neither suspect can be assured of a better outcome.

A much more appealing strategy for both suspects would be not to confess, as long as each could be sure that the partner would also remain silent. But without being able to communicate to coordinate a joint strategy, much less to make a binding agreement, this strategy could backfire and one could end up with 10 years in prison. The reason is that the better outcome than (-5,-5) for *both* players, (-1,-1), is unstable, for there is always the temptation for one player to double-cross his or her partner and turn state's evidence to achieve the very best outcome of being set free (0). Not only is there this temptation, but if one player thinks that the other might try to make a sucker out of him or her, there is no recourse but to confess. Thus, both suspects' strategies of confession *strictly dominate*—are preferable whatever strategy the other player chooses (Section 2.2)—their strategies of not confessing, though the choice of the former strategy by both suspects results in a relatively undesirable 5 years in prison for each.

The dilemma lies in the fact that when both suspects play it safe by choosing their dominant strategies of confession, they end up worse off [(-5,-5)] than had they trusted each other and both not confessed [(-1,-1)]. Since neither player has an incentive to depart from (-5,-5), not only is this outcome Pareto-inferior (Section 6.4), but it is also a Nash equilibrium (Section 6.6). Moreover, it is the only Nash equilibrium: at least one player can do immediately better by departing unilaterally from the other three outcomes [including (-1,-1)].

Now consider prisoners' dilemma as a sequential game, as defined by the four rules of play given in Section 6.6. The initial-outcome matrix, based on ordinal rankings rather than the Figure 7.1 payoffs, is shown in Figure 7.2, with the game tree of sequential moves from (3,3) given below the matrices of initial and final outcomes. Note that the ranking of initial outcomes is the same as that of chicken (Figure 6.2), except for the interchange of ranks of the two worst outcomes (1 and 2) of each player. The two games are also *symmetrical:* both players rank the outcomes the same along the main diagonal, and the ranks of the off-diagonal outcomes are mirror images.

An analysis of the Figure 7.2 game tree demonstrates that neither player, thinking ahead, would be motivated to depart from (3,3), because if one did [say, row moved to (4,1)], the other player (column) would

Figure 7.2 Prisoners' Dilemma as a Sequential Game

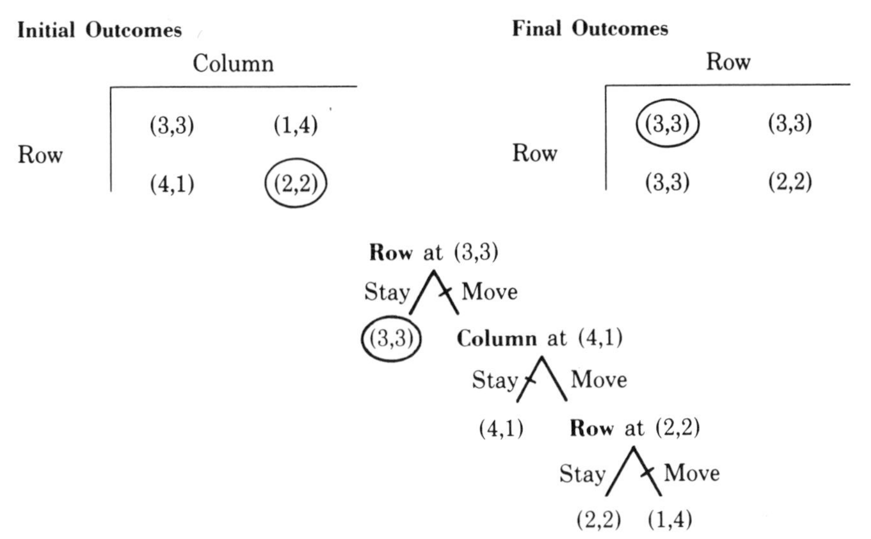

Key: (x,y) = (row, column)
4 = best; 3 = next best; 2 = next worst; 1 = worst
Circled matrix outcomes are dominant-strategy Nash equilibria; circled outcome in game tree is rational outcome (following path of uncut branches, starting at top).

countermove [to (2,2)], where subsequent moves would terminate.

This is so because if row moved from (2,2) to (1,4), column would stay in order to implement his or her best outcome. But since row can anticipate that the process would end up at (2,2) if he or she were to depart initially from (3,3), the players would have no incentive to depart from (3,3) in the first place. Thereby (3,3) is the rational outcome when prisoners' dilemma is played according to the sequential rules.

The two-sided analysis I illustrated in the case of chicken in Section 6.6, when applied to prisoners' dilemma, shows that from initial outcomes (4,1) and (1,4) the process would move to final outcome (3,3). At (4,1), for example, it would be in *both* players' interest that row move to (3,3) before column moves to (2,2), where the process would stop, as indicated in the previous paragraph. Because there is no incentive for row to move to (1,4) or column to (4,1) from (2,2), that outcome, like (3,3), is stable in a nonmyopic sense.

Thus, were (2,2) initially chosen by the players, it would be the final outcome, whereas all other outcomes would be transformed into (3,3). Altogether, the final-outcome matrix of prisoners' dilemma given in Figure 7.2 shows only the upper-left (3,3) Nash equilibrium to be the product of dominant strategy choices by the players—and presumably the outcome that would be chosen by rational players. Note that it coincides with the (3,3) outcome in the initial-outcome matrix; the other two (3,3) outcomes in the final-outcome matrix are Nash equilibria but are not associated with the players' dominant strategies.

Thus, as in chicken, the sequential rules lead to the cooperative (3,3) outcome in prisoners' dilemma; but in prisoners' dilemma the implementation of this outcome does not depend on one player's being stronger and quicker than the other. Not only does the dominance of strategies associated with (3,3) in the final-outcome matrix require no special assumptions, but also the dominance of the players' other strategies, associated with (2,2) in the initial-outcome matrix, is reversed.

Thereby, starting at (3,3), it is not advantageous for either player, looking ahead, to switch strategies. If the status quo is the (2,2) outcome, however, this outcome, because it is nonmyopically (as well as myopically) stable, would appear difficult to shift from in order to reach the (3,3) cooperative outcome.

One might think that if prisoners' dilemma were played repeatedly, perspicacious players could in effect communicate with each other by establishing a pattern of previous choices that would reward the choice of the cooperative strategy. But if the game ends after n plays, it clearly does not pay to play cooperatively on the final round: with no plays to follow, the players are in effect in the same position they would be in if they played the game only once. If there is no point in trying to induce a cooperative response on the nth round, such behavior on the (n − 1)st round would also be to no avail since its effect could extend only to the nth round, where cooperative behavior has already been ruled out. Carrying this reasoning successively backward, it follows that players should not choose the cooperative strategy on *any* plays of the game.

This reasoning is logically impeccable, but recent work by Axelrod suggests that cooperation in repeated play of prisoners' dilemma may not be as hopeless as the previous argument makes it sound—at least when players do not know when play will terminate (it was assumed in the preceding paragraph that the game had a fixed and known number of rounds).[7] Axelrod found that in computer tournament play of prisoners' dilemma, whereby many programs giving strategies for play were matched against each other in a series of encounters, "tit-for-tat" did better than any other program. That is, a player starts out by cooperating ("do not confess" in the previous scenario), but if the other player does not cooperate initially, the first player retaliates on the next round with noncooperation and imitates the other's previous-round behavior in all subsequent play.

The tit-for-tat player does better on average than those who never cooperate or who choose most other strategies. (Nonetheless, there is no unequivocally best strategy in tournament play—it depends on the strategies of one's opponents.) Generally speaking, it pays to be nice (begin by cooperating), provocable (be ready to retaliate quickly if provoked), and forgiving (return to cooperating as soon as the other player does).

Axelrod makes a strong case for the social relevance of his findings, but because his model assumes repeated play against different opponents, its relevance to play against one continuing opponent, as in superpower conflict (Section 7.3), seems limited. Many social-psychological experiments have been conducted on numerous human subjects to measure the effects of various parameters on repeated play of prisoners' dilemma; these experiments, like Axelrod's work, suggest the emergence of cooperative behavior under certain conditions.[8] Yet in one-shot or episodic play of prisoners' dilemma, it is hard to see how a player could escape the dilemma without taking substantial risks.

It is legitimate to ask at this point what relevance prisoners' dilemma has for the study of politics. If the seemingly paradoxical features of strategic choice for players in this game were limited only to hypothetical and artificial situations of the kind faced by the two criminal suspects in the example, one could easily pass this game off as an interesting curiosity but not of significance to real problems of collective choice. In fact, however, instances of prisoners' dilemma have been identified in such diverse areas as agriculture, business, and law, wherein all players can benefit from cooperating but each has a dominant strategy of not cooperating. As suggested in Section 7.1, the problem prisoners' dilemma poses lies at the heart of a theory of the state: citizens can benefit, say, from not paying taxes that support a strong government, but all are hurt more when anarchy takes over because of a weak or nonexistent government. Political philosophers at least since Hobbes have used the anarchy of a stateless society to justify the need for an enforceable social contract and the creation of government, by coercion if necessary, to circumvent the prisoners' dilemma underlying the transfer of authority to a state.[9]

The standard example of prisoners' dilemma in international relations involves an arms race in which two competing nations may either continue the arms race or desist. If both nations desist, they can devote the resources they would otherwise spend on armaments to socially useful projects, which presumably would make both better off and still preserve a balance of power between them. If both arm, both will be worse off for their outlays on socially useless weapons systems and comparatively no stronger militarily. If one nation arms and the other does not, the nation that arms will develop the military superiority to defeat its adversary and thereby realize its best outcome; the nation that disarms will relinquish control over its fate and thereby suffer its worst outcome.

Variants of this example relate to whether to adhere to the conditions of a treaty, cheat on weapons-inspection agreements, and so on, where it is understood that treaties or agreements are not completely enforceable. Central to all these examples is the question of trust: Under what conditions will the players be likely to trust each other sufficiently to risk adopting their cooperative but unstable strategies?

7.3 The Superpower Arms Race as a Prisoners' Dilemma

The nuclear arms race between the United States and the Soviet Union has proved to be one of the most intractable contemporary world problems.[10] Its intractability stems not from the awesome amounts both sides have expended on arms, nor from the millions of lives at stake should the arms race culminate in a nuclear war. Although these facts help to explain why the arms race looms so large in our lives, they do not explain why this race has proved so difficult to slow down.

Several hypotheses for the persistence of the arms race have been advanced: (1) the military-industrial complex in each country is said to hold sway over major policy decisions;[11] (2) the economy of the United States, and perhaps that of the Soviet Union, is alleged to require major military expenditures to avoid recessions or even depressions;[12] (3) the dynamic nature of an arms race is thought to require that each side match or exceed the expenditures of the other side;[13] and (4) where moves toward disarmament are observed, they are claimed to be no more than an elaborate fraud by which the superpowers deceive the rest of the world so that they can maintain their hegemony.[14]

Other purported explanations of the superpower arms race could be cited, but it is not my purpose to catalog or criticize them.[15] For the most part, they are ad hoc, single-factor explanations—sometimes colored by ideological considerations—that are not embedded in a general model that weighs benefits and costs to decision makers in the arms race.

A different kind of explanation of the arms race focuses on the observation that the benefits and costs to each nation are dependent on what *both* nations do; hence, the arms race may be formulated as a game. The usual candidate is prisoners' dilemma, but of course any arms-race model that assumes that nations as players have only two strategies, which lead to well-defined payoffs, is a drastic oversimplification. On the other hand, this simplified model has the advantage that it exhibits in a strikingly elegant way an explanation of the fundamental intractability of the arms race, based only on the consequences of rational behavior by the participants.

Not only shall I attempt to explain the persistence of the arms race, but, shifting to a normative perspective, I shall also investigate a possible solution to the arms race, based on the prisoners' dilemma model, which posits a sequence of moves by the superpowers that may lay the basis for future cooperation and thereby lead to arms-control agreements. Consequences of this sequence are investigated when each side (1) possesses an ability to detect what the other side is doing with a specified probability and (2) pursues a tit-for-tat policy of conditional cooperation—cooperates only if it can determine that the other side is also cooperating. I shall outline when conditional cooperation between the superpowers is rational and, therefore, likely to occur.

It is important to realize that because this analysis is based on a simplified prisoners' dilemma model, it shares the limitations of that model. However, I believe that the probabilistic framework in which the model is embedded and the conditionally cooperative scenario posited add sufficient realism to throw light on present and possible future developments in the arms race. Indeed, the qualitative conclusions about the benefits of conditional cooperation derived from the model suggest some consequences of extended play of prisoners' dilemma—and, by implication, moves in the arms race—that may bode well for long-run cooperation between the superpowers.

A symbolic representation of the payoff matrix of prisoners' dilemma is given in Figure 7.3, along with a game tree that will be described shortly. Interpreted in the context of the arms race between the superpowers, which I call row and column, each has a choice between two strategies, disarm (D) and arm (A). The choice of a strategy by each superpower results in outcome (r_i,c_j), where the subscripts indicate the rankings of the payoff pairs by each player.

As a model of the arms race, this game may be interpreted as follows: each player has a dominant strategy of arming, which leads to the Pareto-inferior Nash equilibrium, (r_2,c_2). As Garthoff put it, "They [the Soviets] would like to have an edge over us [at (r_1,c_4) if they are column], just as we would like to have an edge over them [at (r_4,c_1) if we are row]." [16] The tragic consequence is that both superpowers, by choosing A, are worse off than if they could somehow reach an arms-control agreement and choose D instead, leading to (r_3,c_3). But this outcome is unstable—not a Nash equilibrium—as I showed in Section 7.2.

To be sure, (r_3,c_3) is a nonmyopic equilibrium, and the logic of the theory of moves would seem to dictate its choice. But recall that this logic also sustains (r_2,c_2) as a nonmypoic equilibrium. If the latter outcome accurately describes the present state of affairs in the superpower arms race, then the theory of moves offers no panacea for breaking away from it. Only if (r_3,c_3) were currently the status quo—instead of (r_2,c_2)—would the theory support the logic of staying at the cooperative D-D outcome in

Figure 7.3 Symbolic Representation of Prisoners' Dilemma as a Model of the Superpower Arms Race

		Column: Disarm (D)	Column: Arm (A)
Row	Disarm (D)	(r_3,c_3)*	(r_1,c_4)
	Arm (A)	(r_4,c_1)	((r_2,c_2))*

Row

Disarm — Arm

$E_R(D) = r_3p + r_1(1 - p)$ $\qquad$ $E_R(A) = r_4(1 - p) + r_2p$

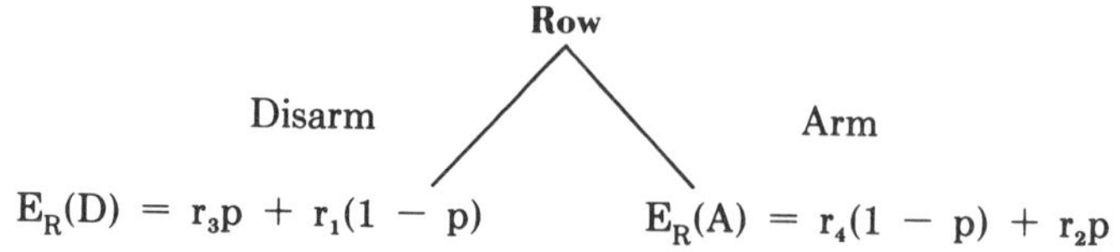

Key: r_4,c_4 = best payoffs; r_3,c_3 = next-best payoffs; r_2,c_2 = next-worst payoffs; r_1,c_1 = worst payoffs
$E_R(D)$ = expected payoff to row for choosing D
$E_R(A)$ = expected payoff to row for choosing A
p = probability that column can correctly detect row's strategy choice
Circled outcome is Nash equilibrium.
Asterisked outcomes are nonmyopic equilibria.

the Figure 7.3 matrix.

Is there another kind of logic that offers more hope? The slow progress that has been made so far in inducing the superpowers to move from (r_2,c_2) to (r_3,c_3) seems to give one little reason to be sanguine. Yet the situation is not hopeless. Because of the recent great increase in the intelligence capabilities of each superpower, each can now make better predictions about the other side's likely choices. If each then follows a tit-for-tat policy of cooperating only when it predicts the other side will cooperate, there is a surprising consequence for the play of prisoners' dilemma: each player has an incentive *not* to choose its second dominant strategy of A if each side's predictions are sufficiently good.

True, if one superpower knows that the other superpower will almost surely choose its second strategy, then that nation should also choose its second strategy to insure against receiving its worst payoff, r_1 or c_1. As a consequence of these choices, the noncooperative outcome, (r_2,c_2), will be chosen.

But now assume that one superpower knows that the other plans—at least initially—to select its first strategy. Then one would ordinarily say that the superpower that discovers this information should exploit it and choose its second strategy, thereby realizing its best payoff of r_4 or c_4.

This tactic will not work, however, given the mutual predictability of choices assumed on the part of both players. For any indications by one player of defecting from its strategy associated with the cooperative but unstable outcome (r_3,c_3) would almost surely be detected by the other player. The other player could then exact retribution—and at the same time prevent its worst outcome from being chosen—by switching to its own noncooperative strategy. Thus, the mutual predictability of strategy choices helps insure against noncooperative choices by *both* players and helps stabilize the cooperative outcome in prisoners' dilemma.

More formally, assume row contemplates choosing either strategy D or strategy A and knows that column can correctly predict its choice with probability p and incorrectly predict its choice with probability 1 − p. Similarly, assume that column, facing the choice between D and A, knows that row can correctly predict its choice with probability q and incorrectly predict its choice with probability 1 − q. Given these probabilities, I shall show that there exists a choice rule that *either* player can adopt to induce the other player to choose its cooperative strategy—based on the expected-payoff criterion—given that the probabilities of correct prediction are sufficiently high.

A *choice rule* is a conditional strategy based on a player's prediction of the strategy choice of the other player. In the calculation to be given shortly, I assume that one player adopts a choice rule of *conditional cooperation:* it will cooperate (that is, choose its first strategy) if it predicts that the other player will also cooperate by choosing its first strategy; otherwise, the player will choose its second (noncooperative) strategy.

Assume that column adopts a choice rule of conditional cooperation. Then if row chooses strategy D, column will correctly predict this choice with probability p and hence will choose strategy D with probability p and strategy A with probability 1 − p. Thus, given conditional cooperation on the part of column, row's *expected payoff* from choosing strategy D, $E_R(D)$—which is the sum of each of its payoffs times its probability of occurring—will be

$$E_R(D) = r_3p + r_1(1 - p)$$

Similarly, row's expected payoff from choosing strategy A, $E_R(A)$, will be

$$E_R(A) = r_4(1 - p) + r_2p$$

$E_R(D)$ will be greater than $E_R(A)$, making the choice of D rational for row in the game tree in Figure 7.3, if

$$r_3p + r_1(1 - p) > r_4(1 - p) + r_2p$$

$$(r_3 - r_2)p > (r_4 - r_1)(1 - p)$$

$$\frac{p}{1 - p} > \frac{r_4 - r_1}{r_3 - r_2}$$

It is apparent that the last inequality is satisfied, and $E_R(D) > E_R(A)$, whenever p (in comparison to $1 - p$) is sufficiently large—that is, whenever p is sufficiently close to 1. If, for example, row's payoffs are $r_4 = 4$, $r_3 = 3$, $r_2 = 2$, and $r_1 = 1$, then the expected payoff of row's first strategy will be greater than that of its second strategy if

$$\frac{p}{1 - p} > \frac{4 - 1}{3 - 2}$$

or $p > 3/4$. That is, by the expected-payoff criterion, row should choose its first (cooperative) strategy if it believes that column can correctly predict its strategy choice with a probability greater than 3/4, given that column responds in a conditionally cooperative manner to its predictions about row's choices. Note that whatever the payoffs consistent with row's ranking of the four outcomes are, p *must* exceed 1/2 because $(r_4 - r_1) > (r_3 - r_2)$.

What happens if column adopts a less benevolent choice rule? Assume, for example, that column always chooses strategy A, whatever it predicts about the strategy of row. In this case, if row now adopts a conditionally cooperative choice rule, it will choose strategy D with probability $1 - q$ and strategy A with probability q. By the symmetry of the game, if the roles of row and column are reversed, one can show, in a manner analogous to the comparison of the expected payoffs of strategies given previously for row, that $E_C(D) > E_C(A)$ if

$$\frac{q}{1 - q} > \frac{c_4 - c_1}{c_3 - c_2}$$

or whenever q (in comparison to $1 - q$) is sufficiently large—that is, whenever q is sufficiently close to 1. Subject to this condition, therefore, column would *not* be well advised always to choose strategy A if row adopts a conditionally cooperative choice rule. Clearly, if both the previous inequalities are satisfied, row and column both do better choosing their cooperative strategies of D to maximize their expected payoffs, given that each player follows a choice rule of conditional cooperation.

7.4 Consequences of Mutual Predictability in Prisoners' Dilemma

So far I have shown that if one player—call it the *leader*—(1) adopts a conditionally cooperative choice rule and (2) can predict the other player's strategy choice with a sufficiently high probability, the other player—call it the *follower*—maximizes its own expected payoff by cooperating also, given that it can detect lies on the part of the leader with a sufficiently high probability. Thereby both players lock into the cooperative outcome, which is myopically unstable in prisoners' dilemma when the players do not have the ability to predict each other's strategy choices.[17]

One question remains, however. Given that the follower maximizes its expected payoff by cooperating when the leader adopts a choice rule of conditional cooperation, how does the follower know when the leader has adopted such a choice rule in the first place? The answer is that it does not (a follower can predict a choice but not a choice rule) unless the leader announces an intention to adopt this choice rule.

For the players to escape the dilemma, therefore, there must be communication between them—specifically, one player (the leader) announces a choice rule to which the other player (the follower) responds. If neither player takes the initiative, nothing can happen; if both players take the initiative simultaneously and announce the choice rule of conditional cooperation, each presumably will await a commitment from the other before committing itself to a move, and again nothing will happen. Should the players simultaneously announce different choice rules, the resulting inconsistencies may lead either to confusion, or, possibly, to an attempt to align the rules or distinguish the roles of leader and follower.[18]

The only clean escape from the dilemma, therefore, occurs when the two players can communicate and take on the distinct roles of leader and follower. Although, strictly speaking, permitting communication turns prisoners' dilemma into a game that is no longer wholly noncooperative, communication alone is not sufficient to resolve the dilemma without mutual predictability. For what is to prevent the leader from lying about its announced intention to cooperate conditionally? And what is to prevent the follower from lying about its announced response of selecting its cooperative strategy?

With mutual predictability players have the insurance that the lies can be detected with probabilities p and q. If these probabilities satisfy the inequalities set forth in Section 7.3, then it pays for the follower to cooperate in the face of a choice rule of conditional cooperation, and for the leader to cooperate by choosing its cooperative strategy, too. Otherwise, the insurance both players have against lying will not be sufficient to make cooperation worth their while; if such is the case, they should, instead, choose their noncooperative dominant strategies. I conclude, therefore, that a mutual ability to predict choices on the part of both players offers them a mutual incentive to choose their cooperative strategies.

It is worth noting that the leader-follower model suggests circumstances under which an *absolutely* enforceable contract will be unnecessary. When the prediction probabilities of the players are sufficiently high (which depends on the payoffs assigned by the players to the outcomes), an agreement to cooperate—reached in leader-follower negotiations before play of the game—can be rendered *enforceable enough* to create a probabilistic kind of equilibrium that stabilizes the cooperative outcome.

The capabilities that the superpowers have of detecting and predicting each other's behavior may be viewed as providing a basis for each to trust the other, for to be trustworthy means, among other things, to be predictable. In effect, predictability, rooted in a player's detection capabilities, helps ensure that a superpower's trust is not misplaced.

To what extent do players in real-world political games think in the terms I have described? This is a difficult question to answer generally, but one specific illustration of this kind of thinking may persuade the reader that it is certainly not unknown in the field of foreign affairs. In describing a highly classified mission, code-named Holystone, that allegedly involved reconnaissance by U.S. submarines inside Soviet waters, one U.S. government official was quoted as saying: "One of the reasons we can have a SALT [Strategic Arms Limitation Talks] agreement is because we know of what the Soviets are doing, and Holystone is an important part of what we know about the Soviet submarine force." [19] Almost a decade later, this judgment about the value of submarine reconnaissance as an element of nuclear deterrence had not changed. According to a naval surveillance officer, surveillance is "part of the arms-reduction talks. The Russians have to know that we're good at tracking them." [20]

Before discussing the role and effects of other kinds of strategic intelligence, I shall extend the probabilistic framework by defining new probabilities, called "detection probabilities" to distinguish them from the "prediction probabilities" defined in Section 7.3. These will be used in a two-stage model that avoids the leader-follower asymmetry of the previous model but reveals new problems with which players who are trying to break out of the noncooperative outcome in prisoners' dilemma must cope.

7.5 The Superpower Arms Race as a Two-Stage Prisoners' Dilemma

Assume that row (R) and column (C) begin the game by both announcing a tit-for-tat policy of conditional cooperation: "I'll cooperate [that is, choose D] if I detect you will; otherwise I won't." Then assume both players, to show their good intentions, initially cooperate and choose D, though neither player assuredly knows that the other player has made this choice. This is the *first stage* of the game.[21]

The *second stage* begins when each player makes a second strategy choice, depending on what it detected its opponent did in the first stage. Assume that R can detect with a certain probability the strategy choice of C and also that C can detect R's strategy choice with a certain probability. Specifically, let

p_R = probability that R can detect C's strategy choice in the first stage

p_C = probability that C can detect R's strategy choice in the first stage

where $0 \leq p_R, p_C \leq 1$.

Consistent with a policy of conditional cooperation, assume that a player chooses D in the second stage if it detects that its opponent chose D in the first stage; otherwise, it chooses A. Hence, R chooses D with probability p_R, and C chooses D with probability p_C—precisely the probabilities that R and C correctly detect each other's assumed choices of D in the first stage. In this manner, the detection probabilities of the players are linked to their (probabilistic) strategy choices in the second stage, though there need be no necessary equivalence; this linkage is an assumption of the model whose implications will now be explored.[22] Specifically, consider the following question: Does a policy of conditional cooperation benefit the players in the second—and perhaps later—stages of the game?

In the first stage, the expected payoff is r_3 for R and c_3 for C, because, by assumption, the cooperative outcome (r_3,c_3) is chosen with probability 1. In the second stage, the expected payoff of R will be

$$E_R = r_3 p_R p_C + r_4(1 - p_R)p_C + r_1 p_R(1 - p_C) + r_2(1 - p_R)(1 - p_C) \tag{7.1}$$

assuming R and C make independent strategy choices based solely on their probabilities of detection.

Thus, for example, the first term on the right-hand side of equation (7.1) indicates that R and C will correctly detect their mutual choices of D in the first stage with probability $p_R p_C$: R will detect that C cooperates with probability p_R, and C will detect that R cooperates with probability p_C. If both players follow a policy of conditional cooperation in the second stage, both will choose D with probability $p_R p_C$; hence, R will obtain a payoff of r_3 with probability $p_R p_C$, and C will obtain a payoff of c_3 with this same probability. The probabilities associated with the three other payoffs of R in equation (7.1) (r_4, r_2, and r_1) can be similarly obtained.

Rearranging terms in equation (7.1) yields

$$E_R = p_C[r_3 p_R + r_4(1 - p_R)] + (1 - p_C)[r_1 p_R + r_2(1 - p_R)] \quad (7.2)$$

Whatever the value of p_R, the first term in brackets on the right-hand side of equation (7.2) will always be greater than the second term in brackets, since $r_3 > r_1$ and $r_4 > r_2$. Therefore, it is in R's interest that p_C be as high as possible (so C will correctly detect cooperation and thereby will also cooperate), and it is in C's interest that p_R be as high as possible.

This conclusion is not surprising. Rearranging terms again in equation (7.1) gives a more curious result:

$$E_R = p_R[r_3 p_C + r_1(1 - p_C)] + (1 - p_R)[r_4 p_C + r_2(1 - p_C)] \quad (7.3)$$

In this case the second term in brackets on the right-hand side of equation (7.3) will always be greater than the first term in brackets, so that it is in R's interest that $(1 - p_R)$ be as high as possible, or that p_R be as low as possible. This is because if R *incorrectly* detects that C chooses A in the first stage and thereby chooses A itself in the second stage, then R obtains a higher expected payoff than if it *correctly* detects cooperation on the part of C. In other words, while it is to R's advantage that p_C be high, it is not to its advantage that p_R be high; it is better to be deceived (that is, incorrectly to detect A on the part of C), giving R an "excuse" not to cooperate and thereby to realize perhaps its best payoff (if C cooperates).

But surely C could anticipate this injurious consequence if it knew that p_R were low. Hence, C should not mechanically subscribe to a policy of conditional cooperation in the second stage unless it is assured that R can predict with a high probability its cooperative choice in the first stage and thereby respond accordingly. A similar conclusion applies to R. Within the present framework that allows p_R and p_C to vary independently, this conflict appears irreconcilable.

It turns out, however, that if one constrains these probabilities by setting them equal to each other, it is in the interest of R and C that *both* p_R and p_C (now a common probability) be as high as possible.[23] One way to equalize these probabilities, which has been proposed in negotiations on a new SALT agreement, is for the superpowers to pool their information so that they both operate from a common (and enlarged) data base.[24] A common data base would presumably have an equalizing effect because it would give neither player an advantage in intelligence—assuming no data were surreptitiously withheld by a player. Alternatively, if "national technical means for verification"—in the terminology of current arms-limitation talks—of both players were equally good, their detection probabilities would also be equal.

In fact, one of the most significant qualitative changes in the nuclear arms race since the early 1960s has been the dramatic rise in the detection capabilities of both sides, which has been principally due to the use of reconnaissance satellites.[25] Indeed, President Lyndon B. Johnson stated that space reconnaissance had saved enough in military expenditures to pay for the entire military and space programs.[26] President Jimmy Carter, in the first public acknowledgment of photo reconnaissance satellites, said that "in the monitoring of arms control agreements they make an immense contribution to the security of all nations."[27]

If the detection capability of either side is destroyed or even threatened, then conditional operation will be rendered unappealing and the prospects of a continuing arms race will be high. On the other hand, if each side's detection capabilities can be ensured or even strengthened—especially through the sharing of data that help render $p_R = p_C$—then further agreements in SALT will appear not only desirable but also rational for both sides.

Stability in the superpower arms race has been based on a policy of deterrence or MAD—the ability and willingness of each side to respond to a possible first strike by the other side—as discussed in Chapter 6. But if MAD has prevented war, it has not led to a significant *diminution* in the arms race, which now seems to depend on the ability of each side to detect cooperation on the part of the other side and to respond to it in kind. Unfortunately, "probably nothing the United States does is more closely held than the techniques and performances of its verification machinery."[28] To promote movement toward an arms-control agreement, it would seem generally in the interest of both the United States and the Soviet Union not only to improve their own detection capabilities but also to abet the detection capabilities of each other.[29]

Naturally one cannot argue as a blanket prescription that all reconnaissance information about weapons systems should be shared. Information that would greatly increase a country's vulnerability to attack may itself create instability by making a preemptive strike more attractive. Thus, the gains of both superpowers from increasing the likelihood of successful mutual disarmament, obtained through a sharing of information that enhances their common detection probability, must be balanced against the possibility that such shared information could increase their vulnerability to a first nuclear strike.

Since I have precluded noncooperation by either superpower in the first stage of the two-stage model, the incentive to strike first is presumed not to exist. Should this incentive exist, however, it would create a fundamental instability that would render the postulated game scenario implausible. At this time, it seems, both superpowers possess substantial second-strike capabilities, stemming principally from the relative invulnerability of their submarine-launched nuclear missiles. Hence, both superpowers have an incentive not to launch first strikes but instead to find some reasonably safe way to move away from a constant repetition of the burdensome (r_2,c_2) outcome. The two-stage model suggests one way this process may be initiated.

It is important to point out factors that may complicate the rational calculations I have suggested based on the expected-payoff criterion. First, the concept of "expected payoff" assumes that the arms race is not viewed as a one-shot affair, but as a multistage game played out in an uncertain environment. Even viewed in these terms, however, there are many possible scenarios, and the consequences of only one have been investigated. It would be useful to investigate other plausible scenarios—perhaps occurring over more than two stages, possibly with allowance made for the discounting of payoffs in later stages[30]—to determine the conditions that make mutual cooperation rational.

7.6 The Effects of Power in Prisoners' Dilemma

In Section 7.2 I indicated that both the cooperative (r_3,c_3) outcome and the noncooperative (r_2,c_2) outcome in prisoners' dilemma are nonmyopic equilibria. It is not difficult to show that if either player has moving power (illustrated in Section 6.4) or threat power (illustrated in Sections 6.2 and 6.4), it can ensure the selection of (r_3,c_3).

I have not yet illustrated the use of *staying power,* which allows the player who possesses it to hold off making a strategy choice until the player without this power makes one. Then moves alternate, according to the theory of moves, starting with the player without staying power, with a "rational termination" condition postulated that prevents cycling, or returning to the initial outcome.[31]

Because prisoners' dilemma contains nonmyopic equilibria, there is no cycling. Since play will therefore terminate before the players return to the initial outcome, it is easy to show that it is rational for the player without staying power to choose its cooperative strategy initially and for the player with staying power to respond by also choosing its cooperative strategy. For if either player is noncooperative, then the terminal outcome, after the alternating rational moves and countermoves by the players, will be (r_2,c_2), which is Pareto-inferior to (r_3,c_3) and hence will not be chosen if the players are rational, according to the theory of moves.

Remarkably, of the 37 (out of 78) 2 x 2 strictly ordinal games that have nonmyopic equilibria, only in chicken and prisoners' dilemma are the (r_3,c_3) nonmyopic equilibria not also Nash.[32] This may in part explain why these two harrowing games have received so much attention in the game-theory literature. Theorists' abiding interest in them seems to stem from an implicit recognition that mutual cooperation can somehow be justified. But how? Nonmyopic equilibria that allow for the possibility of sequential moves and countermoves in dynamic play offer one justification.

The Figure 6.1 Polish game illustrates the case when (threat) power is *effective,* or makes a difference—depending on who possesses it—on what outcome is implemented. By contrast, the exercise of power (threat and moving) in the Figure 6.3 Cuban missile crisis game was ineffective because whichever player possessed the power, the same (3,3) outcome would have been induced. Similarly, power is ineffective in prisoners' dilemma since (r_3,c_3) is the moving, staying, and threat power outcome in this game, whichever player (if any) possesses such power.

Games in which power is ineffective probably better lend themselves to amicable solutions than games in which power is effective—given that one player possesses such power and can thereby ensure implementation of the (same) outcome that each player's power guarantees. On the other hand, if one player's power enables it to implement an outcome better for itself than its adversary's power enables it to implement, then there will be good reason for the players to compete for influence.

Unfortunately in the superpower arms race, neither player probably has an overriding ability to continue moving when the other player must eventually stop (moving power), to delay making an initial choice significantly longer (staying power), or to threaten the other into submission (threat power). Consequently, with both superpowers currently stuck at the Pareto-inferior noncooperative outcome, neither can readily effect a switch to the Pareto-superior cooperative outcome. Instead, one hopes that a policy of conditional cooperation that is tied to the growing detection/prediction capabilities of the two sides will make a gradual, if not a dramatic, shift possible.

There is no lack of drama in the second case of a trap to be discussed. The iron grip it had on the participants held the American people in thrall for many months.

7.7 The White House Tapes Case: The Players and Their Preferences

Richard Nixon is the only person ever to have resigned the U.S. presidency. The most immediate cause of his resignation was a decision by the U.S. Supreme Court ordering the release of certain White House tapes.

On July 24, 1974, the Court unanimously ruled to end Nixon's efforts to withhold the tapes from Special Prosecutor Leon Jaworski in the so-called Watergate cover-up case. That same day the president, through his attorney, James St. Clair, announced his compliance with the ruling—he would release the White House tapes the special prosecutor sought. Fifteen days later the Nixon presidency ended in ruins, a direct result of the Court's action. This case is of special interest because optimal strategies in the game played over the release of the tapes, as I shall reconstruct it, led to a paradoxical consequence—a trap—that seems to challenge the rationality of the players' choices.[33]

The immediate history of the White House tapes decision began on March 1, 1974, when a grand jury indicted seven former White House and campaign aides for attempting to cover up the Watergate scandal (*United States v. Mitchell et al.*). On April 16 the special prosecutor petitioned Judge John Sirica to subpoena tapes and documents of 64 presidential conversations with John Dean, Robert Haldeman, John Erlichman, and Charles Colson; the subpoena was issued on April 18.

On May 1 St. Clair announced that the president would refuse to submit the subpoenaed materials, and St. Clair sought an order quashing the subpoena. After hearing arguments, Judge Sirica confirmed the subpoena order on May 20. On May 24, St. Clair filed an appeal in the court of appeals, which, it seemed, would probably result in the postponement of the cover-up trial.

The prosecutors moved quickly to prevent delay. On the day the appeal by St. Clair was filed in the court of appeals, Leon Jaworski, using a seldom-invoked procedure, went to the Supreme Court and sought a writ of certiorari before judgment that would leapfrog the appeals process. Citing the imminent cover-up trial date, Jaworski also noted the necessity to settle expeditiously an issue that was paralyzing the government. He requested the Court not only to issue the writ but also, because of the "imperative public importance" of the case, to stay in session into the summer.[34] This way the case could be decided in sufficient time that the tapes could be used as evidence at the trial—should Judge Sirica's ruling be upheld. The Supreme Court agreed on May 31 and heard oral arguments on July 8.

When the justices went into conference on July 9, each of the eight who were to consider the case had basically two choices—decide for or decide against the president. (Associate Justice William Rehnquist withdrew from the case, evidently because of his previous service at the Justice Department under John Mitchell, though he never publicly stated a reason for disqualifying himself.) It appears from the available record that six of the justices reached an early consensus against the president on all three of the major issues: (1) whether the Court had jurisdiction in the case—standing to sue—since Jaworski was an employee of the executive branch; (2) whether executive privilege was absolute; and (3) whether Jaworski had demonstrated a sufficient need for the subpoenaed materials.

Justices Warren E. Burger and Harry A. Blackmun, while concurring with the majority on limiting executive privilege, believed that the special prosecutor lacked legal standing to sue the president. For this reason, it appears, they voted originally against granting the case certiorari.[35]

Justices Burger and Blackmun are conceived of as one player. This is so because it is almost axiomatic that Blackmun votes with Burger: in the first five terms (1970-1974) that Burger and Blackmun served together on the Court, they agreed on 602 of the 721 cases they both heard (83.5 percent), which is the highest agreement level of any pair of justices who served over these five terms.[36] They are referred to as the "Minnesota Twins" by the Supreme Court staff. As deliberations developed, Burger and Blackmun had a choice of two strategies:

1. To decide against the president, joining the other six justices to create a unanimous decision
2. To decide for the president, forming a minority to create a six-to-two "weak" decision

President Nixon's possible response to an adverse Supreme Court ruling was long a matter of doubt—and one that, it will be argued here, Burger and Blackmun could not afford to ignore. On July 26, 1973, White House Deputy Press Secretary Gerald Warren stated that President Nixon would abide by a "definitive decision of the highest court." Nixon, at a news conference on August 22, 1973, endorsed the Warren formulation, but neither he nor White House spokesmen would expand on the original statement.[37] These statements were made in reference to the president's refusal to obey a subpoena from the first special prosecutor, Archibald Cox, for nine White House tapes.

That case never reached the Supreme Court. The court of appeals ruled against the president, who, after a delay of 11 days, agreed to submit the tapes—but not before he had dismissed Cox. The question of what "definitive" meant then became moot.

The issue arose again on May 24, 1974, when Jaworski filed his appeal with the Supreme Court. On July 9 St. Clair made it clear that the president was keeping open the "option" of defying the Court. The question of compliance, he stated, "has not yet been decided." [38] Since the expectation at the time was that the Court would rule against the president,[39] President Nixon had two strategies:

1. Comply with an adverse Court ruling
2. Defy an adverse Court ruling

Several factors help to explain President Nixon's refusal to make a definite commitment concerning his response to a Court decision. If he stated that he would not comply, his statement might be used as a ground for impeachment. If he stated that he would comply, then the House Judiciary Committee might argue that the president would either have to comply with its subpoenas, too, or be impeached.[40]

More important, though, the president's refusal to assure his compliance with an adverse decision was designed to threaten the Court and lead the justices to render either a favorable decision or, at worst, a closely divided adverse split decision that he could claim was insufficiently "definitive" for a matter of this magnitude. Evans and Novak reported at the time, "The refusal of St. Clair to say Nixon would obey an adverse decision has disturbed the judicial branch from the high court on down."[41]

If the president's intent were to threaten the Court, the threat backfired. Why? To explain why Justices Burger and Blackmun departed from their apparent personal preferences and eventually sided with the Court majority, I shall next describe the game they and Nixon played.

The probable outcomes of the four possible strategy choices of the two players are presented in the outcome matrix in Figure 7.4. As a

Figure 7.4 Outcome Matrix of White House Tapes Game

		Nixon	
		Comply with Court	Defy Court
Burger and Blackmun	Decide for president; create a "weak" 6-2 decision	I. Constitutional crisis averted; Nixon not impeached for non-compliance; majority-rule principle preserved	II. Constitutional crisis; Nixon impeached but conviction uncertain
	Decide against president; create a unanimous 8-0 decision	IV. Constitutional crisis averted; Nixon not impeached for non-compliance; majority-rule principle possibly weakend	III. Constitutional crisis; Nixon impeached and conviction certain

justification for these outcomes, first consider the consequences associated with Nixon's defiance of an adverse Supreme Court ruling.

Unquestionably, if the president defied the Court, his defiance would represent a direct assault on the Supreme Court's constitutional place as the "principal source and final authority of constitutional interpretation" and thereby threaten the very structure of the American political system.[42] Indeed, it seems highly probable that Nixon would have plunged the country into its deepest constitutional crisis since the Civil War. No previous president had ever explicitly defied an order of the Supreme Court, though such action had apparently been contemplated.[43]

At the time of the decision in *United States v. Nixon,* it appeared that the result of presidential defiance would be impeachment by the House of Representatives on the ground of withholding evidence from the special prosecutor or violation of the principle of separation of powers. While the outcome in the Senate was less certain than that in the House, a unanimous adverse decision by a Court that included three conservative Nixon appointees (Burger, Blackmun, and Lewis F. Powell, Jr.) would preempt charges that the president was the victim of what presidential counselor Dean Burch called a "partisan lynch mob." [44] (I do not include Powell as a Court player because he originally favored granting certiorari; also, he "demonstrated the highest level of independence within the Nixon Bloc" and has been described as "one of the least predictable of the eight and most flexible of the Nixon appointees.")[45] Indeed, on the day of the decision St. Clair warned the president that he would be surely impeached and swiftly convicted if he were to defy the unanimous ruling of the Court.[46]

On the other hand, Jaworski believed that "if the vote against [the president] was close he would go on television and tell the people that the presidency should not be impaired by a divided Court." [47] A "weak" decision from which at least some of the more conservative Nixon appointees dissented would also allow the president to continue his "one-third plus one" strategy in the Senate to avoid conviction and removal from office (by a two-thirds or greater majority).

Consider now the consequences associated with Nixon's compliance with an adverse Supreme Court decision. Clearly, compliance would avert a constitutional crisis, and Nixon would thereby avoid immediate impeachment in the House for not complying with the Court. However, compliance posed problems for the president; he had reason to believe that the subpoenaed materials, if released, would prove damaging and might even lead to his eventual impeachment by the House. In fact, upon hearing of the Court's decision, Nixon, who was at his San Clemente, California, home, telephoned White House special counsel Fred Buzhardt in Washington. "There may be some problems with the

June 23 tape," Nixon said.[48]

Although the revelation of this tape ultimately forced his resignation, Nixon apparently did not fully realize at the time the incriminating nature of the recorded conversations. In *The Final Days,* Woodward and Bernstein report that Buzhardt felt that the tape was "devastating." Nixon, on the other hand, felt that Buzhardt was "overreacting," that it was "not that bad."[49] Even as late as August 5, in his statement accompanying the public release of the tape transcripts, Nixon reflected his mixed evaluation of the tape's impact: "I recognize that this additional material I am now furnishing may further damage my case.... I am firmly convinced that the record, in its entirety, does not justify the extreme step of impeachment and removal from office."[50]

Compliance—or, more accurately, the announcement of compliance—would allow the president to fall back on his long-used strategy of delay, though it would not necessarily remove the threat of impeachment and ultimate conviction, especially if the Court were unanimous in its judgment. For Justices Burger and Blackmun, who had voted originally against granting the case review, supporting and enlarging the majority (possibly against their convictions) to counter a presumed threat to the Court's authority might possibly weaken the majority-rule principle that *any* majority is sufficient for a decision.[51] But voting their convictions would be hazardous should the president use a divided decision as a pretext to defy the Court.

I shall now attempt to combine these conflicting considerations into a ranking of the four outcomes by the two players. As in previous games, I shall not try to assess how much each player preferred one outcome over another.

Clearly, President Nixon preferred the *risk* of conviction and removal to its virtual certainty. Thus, the president would prefer to defy a weak decision (outcome II in Figure 7.4) than to defy a unanimous decision (III), which I indicate by the (partial) preference scale (II,III). For the same reason, he would prefer to comply with any adverse decision (I or IV) than to defy a unanimous decision (III)—his worst outcome—so (I-IV,III), where the hyphen indicates indifference between I and IV.

Defying a weak decision (II) is considered preferable to complying with any adverse decision (I or IV), for such defiance would preclude the release of potentially devastating evidence and at the same time present Nixon with the possibility of avoiding conviction and removal for noncompliance; hence (II,I-IV). Between the two compliance outcomes (I and IV), I assume that the president "preferred" to comply with a weak decision (I) over a unanimous decision (IV), so (I,IV). A weak decision with some justices dissenting would leave the issue confused and subject to interpretation; a weak decision would also leave room to maneuver for partial compliance.[52] Putting the partial preference scales together, the president's presumed ranking of the four outcomes is: II preferable to I preferable to IV preferable to III, or (II,I,IV,III). Given these rankings by the president, what are the corresponding rankings of Burger and Blackmun?

Although it was previously suggested that Burger and Blackmun would have preferred to decide for the president on at least one of the strictly legal questions (standing to sue by the special prosecutor), there is no doubt that the justices believed that compliance by the president with any adverse Court ruling (I or IV) would be preferable to defiance (II or III); hence, their partial preference scale is (I-IV,II-III). Indeed, in the Court's opinion, which Burger drafted, the chief justice quoted Chief Justice John Marshall in *Marbury v. Madison* (1803): "It is emphatically the province and duty of the Judicial department to say what the law is."

It also seems reasonable to assume that if the president complied, the justices would prefer to decide for him (I) rather than against him (IV); hence, (I,IV). After all, the notion that the Court must be unanimous or close to it to make a decision credible, and thereby induce compliance, is an undesirable restriction on the Court's authority and might establish an unhealthy precedent. Finally, I assume that the justices "preferred" the president to defy a unanimous decision (III) rather than to defy a weak decision (II)—on which his chances of eventual success would be higher—so (III,II).

Putting the partial preference scales together, the justices' presumed ranking of the four outcomes is: I preferable to IV preferable to III preferable to II, or (I,IV,III,II). The resulting rankings of the four outcomes in the Figure 7.4 outcome matrix are shown in the Figure 7.5 payoff matrix.

Figure 7.5 Payoff Matrix of White House Tapes Game

		Nixon	
		Comply with Court	Defy Court
Burger and Blackman	Decide for president	I. (4,3)	II. (1,4)
	Decide against president	IV. (3,2)	III. (2,1)

Key: (x,y) = (Burger and Blackmun, Nixon)
4 = best; 3 = next best; 2 = next worst; 1 = worst

7.8 The White House Tapes Case: The Trap

Because the players in the White House tapes game did not make simultaneous choices in ignorance of each other, the payoff matrix in Figure 7.5 does not provide an accurate representation of this game. As in the Esther game (Section 2.2), it is the most convenient way to describe player preferences.

In fact Burger and Blackmun—and the rest of the Court—acted first. Only then did Nixon have to make a strategy choice, as represented in the revised 2 x 4 payoff matrix shown in Figure 7.6, wherein Burger and Blackmun have two strategies while Nixon has four. This is so because each of Nixon's two original strategies, "comply" and "defy," are contingent on what Burger and Blackmun decide ("decide for president," "decide against president"), which yields (2)(2) = 4 strategies for the president.

Now consider the White House tapes game shown in Figure 7.6. It is easy to see that D if F, C if A is a dominant strategy for Nixon: it yields payoffs as good as, and in at least one case better than, the payoffs yielded by any of his other three strategies, whatever the strategy choice of Burger and Blackmun (F or A).

Given this unconditionally best strategy choice on the part of Nixon, it is reasonable to assume that Burger and Blackmun will anticipate

Figure 7.6 Revised Payoff Matrix of White House Tapes Game

		Nixon			
		Comply (C) regardless	Defy (D) regardless	C if F, D if A	D if F, C if A
Burger and Blackmun	Decide for president (F)	(4,3)	(1,4)	(4,3)	(1,4)
	Decide against president (A)	(3,2)	(2,1)	(2,1)	(3,2)
					↑ Dominant strategy for Nixon

Key: (x,y) = (Burger and Blackmun, Nixon)
4 = best; 3 = next best; 2 = next worst; 1 = worst
Circled outcome is Nash equilibrium.

Nixon's choice, assuming they (as well as Nixon) have complete information about the Figure 7.6 revised payoff matrix—or at least an intuitive understanding of the strategic situation as presented here. To maximize *their* payoff, Burger and Blackmun will choose the strategy that yields for them the highest payoff in the column associated with Nixon's dominant strategy. Since 3 is better than 1 for Burger and Blackmun in this column, one would therefore expect that they would choose their strategy A.

As already indicated, the Supreme Court did decide unanimously against President Nixon. Nixon was reportedly shocked by the Court's ruling, feeling himself "sold out" by his three appointees, Chief Justice Burger and Associate Justices Blackmun and Powell. Charles Colson claims that the president counted on all three justices. Others say he was certain of Burger and Blackmun. When he learned of the decision, Nixon used foul ("expletive-deleted") language to describe Burger. The president could not believe that the Court's ruling had been unanimous. "Significantly, the President's greatest fury seems to have been directed not at the decision itself but at the three Justices who 'deserted' him." [53]

In any event, the decision was unanimous, with no dissenting or concurring opinions. "It was the Court's seamless unity which made defiance so difficult." [54] Eight hours after the decision was handed down, the president, through St. Clair, announced his compliance with the decision "in all respects."

In summary, the game-theoretic analysis seems to explain well, in terms of the foregoing reconstruction of the players' strategies and preferences for outcomes, why the players acted as they did. Yet not only is the outcome that occurred not the most desirable one from the viewpoint of either player, but also both players might have done better had the president been a little more reassuring.

It is worth noting that the lower right (3,2) payoff is the unique Nash equilibrium in the Figure 7.6 matrix: once chosen by both players, neither player has an incentive to depart unilaterally from it because he can do no better, and perhaps worse, if he does. Yet, paradoxically, *both* players could have done better if they had chosen strategies associated with either of the two (4,3) payoffs in the Figure 7.6 matrix. The outcomes that yield these payoffs both involve the choice by Burger and Blackmun of deciding for the president, and the choice by Nixon of compliance. The president can "arrive at" this choice by selecting either comply (C) regardless or C if F, D if A in the Figure 7.6 matrix.

Unfortunately for the players, however, neither of the outcomes that yield (4,3) as a payoff is in equilibrium: Nixon in each case has an incentive to depart unilaterally from the strategies associated with (4,3) to try to bring about his best outcome, (1,4).[55] Not only are the (4,3) outcomes not in equilibrium but also Nixon's two strategies associated with these outcomes are dominated by his (dominant) strategy, D if F, C if A.

For these reasons, therefore, it is hard to see how both players could have done better, even though the opportunity existed. Only if Burger and Blackmun had believed that their dissent would not trigger presidential defiance could they have voted their (presumed) convictions with greater equanimity. The public record shows that Burger and Blackmun never received any assurance that the president would comply if the Court split. Quite the contrary: Nixon and his spokesmen, as indicated earlier, continually held out the possibility of defying a Supreme Court decision that was not "definitive." Thus, Burger and Blackmun had no choice—despite their disagreement with some arguments of the special prosecutor—but to decide against the president. Thereby the Supreme Court decision was rendered unanimous and both players in the White House tapes game lost out, in a sense, on greater payoffs that—at least in principle—were attainable.

The public probably gained from this "noncooperative" solution, however. If one identifies the public with the special prosecutor in the White House tapes case, it seems likely that the special prosecutor, who set up the game that I have described though he was not himself a player, would rank the outcome that actually occurred as the best of the four possible outcomes. This is certainly a reasonable inference from Jaworski's remarks immediately after the Court decision: "I feel right good over what happened. We can move ahead now. . . . I'm especially pleased it was a unanimous decision. It doesn't leave any doubt in anyone's mind." [56]

It seems worth pointing out that a variety of bizarre motives ("need to fail," "death wish") and personality traits ("self-destructive") have been attributed to Richard Nixon. The analysis here, however, suggests that his stance in the White House tapes case, which pushed his confrontation with the special prosecutor and then the Supreme Court beyond the point of no return, was not at all strange. Rather, Nixon was simply caught up in an intractable game that perhaps with greater prescience he could have avoided.

Political leaders of all different stripes have similar failings and lack of foresight. The consequences of these failings, I believe, can be well understood within the game-theoretic framework used in this and previous chapters. In Nixon's case, his reason for resigning in the end seems eminently sensible: "I no longer have a strong enough political base [to complete the term of office]." [57] This rationale is as good as his reason for obeying the Supreme Court's edict—that he could not do better by defiance. Thus, even a political leader as complex and enigmatic as Nixon seems at root a rational actor.

7.9 Conclusions

In this chapter I have focused on issues of social cooperation and conflict, examining why structural conditions may make cooperation difficult to achieve. I showed that in the game of prisoners' dilemma it is individually rational for players not to cooperate, despite the fact that through cooperation all players can achieve the collectively rational outcome that results in higher payoffs for all. But because the strategies associated with this compromise outcome are not in equilibrium, the players are tempted to defect from it to try to obtain their best outcomes, which unfortunately are the worst outcomes for the other players.

The arms race between the superpowers was conceptualized as a prisoners' dilemma, with the additional property that each player can predict or detect cooperation or noncooperation on the part of the other player with a specified probability. The first model required that one player (the leader) take the initiative and propose to the other player (the follower) a choice rule of conditional cooperation. It did not, however, require a binding and enforceable contract between the two players, nor did it require that the players rely solely on good will and mutual trust. Instead, the analysis suggested that mutual predictability of choices renders the cooperative strategies less risky for both players in this vexing game.

If such predictability obtains, then a contract is unnecessary, for violations will be predictable with a high probability before play of the game, and appropriate sanctions can be applied to the violator in the play of the game. But because such retribution works to the disadvantage of both players, the ability of both players to predict each other's choices serves also to reinforce trustworthy behavior, which is not encouraged in prisoners' dilemma without mutual predictability.

In the second model, consequences of the following two-stage scenario were investigated: both players cooperate in the first stage; each player, knowing the other player's detection probability, follows a policy of conditional cooperation in the second stage. Although the players are not motivated to cooperate—even conditionally—when the detection probabilities are not equal, when they are equal, both players benefit. The power of one player to implement the cooperative outcome might also enable the players to escape the noncooperative trap, but in the real world neither superpower seems to have the capability to enforce the cooperative outcome in this game.

I believe that the kind of mutual predictability and detectability assumed in the models has given impetus to negotiations between the superpowers in SALT and other forums. With each superpower's reconnaissance satellites and other means of intelligence able to detect substantial violations quickly—at least for certain kinds of weapons systems—the abrogation of an agreement by one party will be known before its consequences prove disastrous to the other party and prevent it from taking appropriate countermeasures. With little to be gained from such a violation and perhaps much to be lost, it is less likely to occur.

In fact, each side may even have an incentive to back arms-control measures for weapons, such as ballistic missiles, that can readily be detected. In this manner, space-age technology has fostered arms-control agreements that—because of the ease with which violations could previously be kept secret—have been so difficult to obtain in the past. However, weapons that are easier to conceal, such as cruise missiles, will undoubtedly continue to plague arms-control negotiators.

The White House tapes game was not a prisoners' dilemma (only one player had a dominant strategy, whereas both do in prisoners' dilemma), but it had some of its earmarks. Although the White House tapes game had unfortunate ramifications for the players, the country as a whole probably avoided a trap—even if its citizens were outraged by Watergate.

Larger games, involving sometimes millions of people who can gain certain benefits without incurring costs, may induce "free riders" (who are analogous to prisoners who defect). Thus, for example, most people gain from consumer and environmental groups, but they need not contribute to them—and in fact have an incentive not to—to enjoy the benefits they bring to the general population. Just as the *public goods* these groups provide are nonexclusionary, so are *public bads,* such as air pollution, whose ill effects bring suffering to all.

The problem of supplying public goods and discouraging public bads is structurally related to prisoners' dilemma-type situations. Players have a dominant strategy of not contributing to the public good, or not preventing the public bad from occurring (e.g., by removing pollution-control devices from their cars), for this action is less costly to them, regardless of whether others act benevolently or not.

Although I have not analyzed such situations, they are traps, also.[58] I chose not to focus on them since essentially two-person games like the superpower arms race throw into bold relief the central dilemma they pose—the divergence of individually and collectively rational outcomes. This divergence is ubiquitous in politics, entrapping actors ranging from individual citizens to nation-states and giving rise to pervasive conflict if no means is found to induce or enforce a cooperative outcome on the players. Sometimes this outcome occurs, as when citizens accede certain rights to a strong government, but frequently no compromise is reached, nor is a solution imposed, and everybody is hurt by unending conflict.

NOTES

1. Lewis F. Richardson, *Arms and Insecurity: A Mathematical Study of the Causes and Origin of War* (Pittsburgh: Boxwood, 1960), p. 12.
2. Konrad Lorenz, *On Aggression,* trans. Marjorie Kerr Wilson (New York: Harcourt, Brace and World, 1966), pp. 237, 243. For a critique of this and related work, see *Man and Aggression,* ed. M. F. Ashley Montagu (New York: Oxford University Press, 1968).
3. Thomas Hobbes, *Leviathan* (1651; New York: Liberal Arts Press, 1958).
4. Jean-Jacques Rousseau, *The Social Contract* (1762; New York: Hafner, 1974), particularly book 1, chap. 7; book 2, chap. 3; and book 4, chap. 8.
5. John G. Cross and Melvin J. Guyer, *Social Traps* (Ann Arbor: University of Michigan Press, 1980).
6. This is implicit in the utilities, which I assume incorporate moral and ethical considerations as well as more utilitarian factors.
7. Robert Axelrod, *The Evolution of Cooperation* (New York: Basic, 1984).
8. The experimental literature on prisoners' dilemma is too vast to cite here. The first major study is Anatol Rapoport and Albert M. Chammah, *Prisoner's Dilemma: A Study in Conflict and Cooperation* (Ann Arbor: University of Michigan Press, 1965).
9. Michael Taylor, *Anarchy and Cooperation* (London: Wiley, 1976).
10. This and the next two sections draw on material in Steven J. Brams, "Newcomb's Problem and Prisoners' Dilemma," *Journal of Conflict Resolution* 19 (December 1975): 596-612; and Steven J. Brams, Morton D. Davis, and Philip D. Straffin, Jr., "The Geometry of the Arms Race," *International Studies Quarterly* 23 (December 1979): 567-588.
11. Carroll W. Pursell, Jr., ed., *The Military-Industrial Complex* (New York: Harper and Row, 1972); Sam C. Sarkesian, ed., *The Military-Industrial Complex: A Reassessment* (Beverly Hills, Calif.: Sage, 1972); and Steven Rosen, ed., *Testing the Theories of the Military-Industrial Complex* (Lexington, Mass.: Heath, 1973).
12. Kenneth E. Boulding, ed., *Peace and the War Industry* (New Brunswick, N.J.: Transaction, 1973); Bernard Udis, ed., *The Economic Consequences of Reduced Military Spending* (Lexington, Mass.: Heath, 1973); and Wassily W. Leontief and Faye Duchin, *Military Spending: Facts and Figures, Worldwide Implications, and Future Outlook* (New York: Oxford University Press, 1983).
13. Richardson, *Arms and Insecurity;* for recent work on "Richardson-type process models," see Dina A. Zinnes and John V. Gillespie, eds., *Mathematical Models in International Relations* (New York: Praeger, 1976); Michael D.

Intriligator and Dagobert L. Brito, "Formal Models of Arms Races," *Journal of Peace Science* 2 (Spring 1976): 77-96; and John V. Gillespie et al., "An Optimal Control Model of Arms Races," *American Political Science Review* 71 (March 1977): 226-244.

14. John W. Spanier and Joseph L. Nogee, *The Politics of Disarmament: A Study in Soviet-American Gamesmanship* (New York: Praeger, 1962); and Alva Myrdal, *The Game of Disarmament: How the United States and Russia Run the Arms Race* (New York: Random House, 1976).
15. A recent assessment of different political-economic explanations can be found in Miroslav Nincic, *The Arms Race: The Political Economy of Military Growth* (New York: Praeger, 1982). For a critical review of five recent books on purported explanations of war—not just superpower conflict—see Urs Luterbacher, "Last Words About War?" *Journal of Conflict Resolution* 28 (March 1984): 165-181. The work most relevant to the rational-choice approach taken here, but one that uses expected-utility models rather than game-theoretic models, is Bruce Bueno de Mesquita, *The War Trap* (New Haven, Conn.: Yale University Press, 1981).
16. Raymond L. Garthoff, "The Role of Nuclear Weapons: Soviet Perceptions," in *Nuclear Negotiations: Reassessing Arms Control Goals in U.S.-Soviet Relations,* ed. Alan F. Neidle (Austin, Texas: Lyndon B. Johnson School of Public Affairs, 1982), pp. 10-11.
17. The nonmyopic stability of this outcome, as shown in Section 7.2, depends on the players' anticipation of future moves; but here, I assume, the players are concerned only with predicting each other's initial strategy choices, not with subsequent moves from the initial outcome.
18. The so-called Stackelberg solution in duopoly theory in economics also distinguishes between a leader and a follower. See John M. Henderson and Richard E. Quandt, *Microeconomic Theory: A Mathematical Approach,* 2d ed. (New York: McGraw-Hill, 1971), pp. 229-231.
19. *New York Times,* May 25, 1975, p. 42.
20. Thomas B. Allen and Norman Polmar, "The Silent Chase: Tracking Soviet Submarines," *New York Times Magazine,* January 1, 1984, p. 14; see also John Tierney, "The Invisible Force," *Science '83,* November 1983, pp. 65-78.
21. Other scenarios are, of course, possible. But these moves seem the most plausible if both players are seriously interested in slowing down the arms race. Empirical evidence in support of the scenario posited can be found in William A. Gamson and André Modigliani, *Untangling the Cold War: A Strategy for Testing Rival Theories* (Boston: Little, Brown, 1971). Further justification for conditional cooperation is offered later in this section, where it is argued that this choice rule's "rationality" depends on maintenance of a second-strike capability that is not compromised by initial cooperation (in the first stage).
22. Mathematically, this assumption says that the conditional probability that R (C) chooses D, given that it detects the choice of D by C (R), is 1; and the conditional probability that R (C) chooses A, given that it detects the choice of A by C (R), is 1. In extensive form, each player chooses either D or A in the first

stage; then, after detecting one or the other choice (in fact, by assumption, D is always chosen, but this choice will not always be detected), each player chooses D or A in the second stage with expected payoffs to be given. This latter choice, unlike that in the first stage, is not postulated to be either D or A but assumed to be rationally based: D if a player detects D, A if it detects A. The "rationality" of this tit-for-tat policy, however, is what I challenge in this section; it is challenged in different form in Steven J. Brams, *Superpower Games: Applying Game Theory to Superpower Conflict* (New Haven, Conn.: Yale University Press, 1985), chap. 4, which is basd on Steven J. Brams and Morton D. Davis, "The Verification Problem in Arms Control: A Game-Theoretic Analysis," in *Interaction and Communication in Global Politics,* ed. Claudio Cioffi-Revilla, Richard L. Merritt, and Dina A. Zinnes (Beverly Hills, Calif.: Sage, 1985). For a critique of the present model as developed in Brams, Davis, and Straffin, "The Geometry of the Arms Race," see the comment by Raymond Dacey, "Detection and Disarmament: A Comment on 'The Geometry of the Arms Race,'" *International Studies Quarterly* 23 (December 1979): 589-598; a response is given in Steven J. Brams, Morton D. Davis, and Philip D. Straffin, Jr., "A Reply to 'Detection and Disarmament,'" *Internatioal Studies Quarterly* 23 (December 1979): 599-600. Further refinements in this framework can be found in Raymond Dacey, "Detection, Inference and the Arms Race," in *Reason and Decision,* Bowling Green Studies in Applied Philosophy, vol. 3, ed. Michael Bradie and Kenneth Sayre (Bowling Green, Ohio: Applied Philosophy Program, Bowling Green State University, 1982), pp. 87-100. For another critique of the prisoners' dilemma model, see R. Harrison Wagner, "The Theory of Games and the Problem of International Cooperation," *American Political Science Review* 77 (June 1983): 330-346; in response, see the comment by Steven J. Brams, Morton D. Davis, and Philip D. Straffin, Jr., "Communications," *American Political Science Review* 78 (June 1984): 495.

23. The derivation is given in Brams, Davis, and Straffin, "The Geometry of the Arms Race."
24. *New York Times,* April 27, 1977, p. A7. That data be collected and verified under U.N. auspices is proposed in Alva Myrdal, "The International Control of Disarmament," *Scientific American,* October 1974, pp. 21-23.
25. Les Aspin, "The Verification of the SALT II Agreement," *Scientific American,* February 1979, pp. 38-45; Lynn R. Sykes and Jack Evernden, "The Verification of a Comprehensive Nuclear Test Ban," *Scientific American,* October 1982, pp. 47-55; and Stephen M. Meyer, "Verification and Risk in Arms Control," *International Security* 8 (Spring 1984): 111-126. For a history of aerial reconnaissance programs since the early 1950s, see Herbert F. York and G. Allen Greb, "Strategic Reconnaissance," *Bulletin of Atomic Scientists,* April 1977, pp. 33-42; and Jeffrey T. Richelson, "The Keyhole Satellite Program," *Journal of Strategic Studies* 7 (June 1984): 121-153.
26. W. F. Biddle, *Weapons, Tecnology, and Arms Control* (New York: Praeger, 1972), p. 252.
27. *Chicago Tribune,* October 2, 1978, p. 2.

28. John Newhouse, *Cold Dawn: The Story of SALT* (New York: Holt, Rinehart and Winston, 1973), p. 14; security aspects of reconnaissance programs are discussed in York and Greb, "Strategic Reconnaissance"; and Meyer, "Verification and Risk in Arms Control."
29. Cooperation between the superpowers may also work to their advantage with respect to third parties. When the Soviet Union alerted the United States to possible preparations by South Africa for a nuclear test in August 1977, both countries allegedly worked together to exert political pressure that apparently forestalled the test. *New York Times,* August 28, 1977, p. 1.
30. On this point, see Taylor, *Anarchy and Cooperation;* and Axelrod, *The Evolution of Cooperation;* the discount rate is related to the termination probability in Stephen J. Majeski, "Arms Race as Prisoner's Dilemma Games," *Mathematical Social Sciences* 7 (June 1984): 253-260.
31. Steven J. Brams and Marek P. Hessel, "Staying Power in 2 x 2 Games," *Theory and Decision* 15 (September 1983): 279-302; and Steven J. Brams, *Superior Beings: If They Exist, How Would We Know? Game-Theoretic Implications of Omniscience, Omnipotence, Immortality, and Incomprehensibility* (New York: Springer-Verlag, 1983), chap. 5.
32. Steven J. Brams and Donald Wittman, "Nonmyopic Equilibria in 2 x 2 Games," *Conflict Management and Peace Science* 6 (Fall 1981): 39-62; a complete listing of the 2 x 2 games of conflict, equilibria they contain, and their moving, staying, and threat power outcomes is given in Brams, *Superior Beings,* Appendix.
33. This and the next section are drawn from Steven J. Brams and Douglas Muzzio, "Game Theory and the White House Tapes Case," *Trial* 13 (May 1977): 48-53; and Brams and Muzzio, "Unanimity in the Supreme Court: A Game-Theoretic Explanation of the Decision in the White House Tapes Case," *Public Choice* 32 (Winter 1977): 67-83. Additional background information can be found in the original articles and in Douglas Muzzio, *Watergate Games* (New York: New York University Press, 1982).
34. J. Anthony Lukas, *Nightmare: The Underside of the Nixon Years* (New York: Viking, 1976), p. 495.
35. Nina Totenberg, "Behind the Marble, Beneath the Robes," *New York Times Magazine,* March 16, 1975, pp. 15ff.
36. Data on case agreement can be found in the November issues of *Harvard Law Review* (1971-1975). On the concurrence of Burger and Blackmun, see *New York Times,* July 1, 1974, p. 10; and Totenberg, "Behind the Marble, Beneath the Robes."
37. *New York Times,* July 25, 1974, p. 22.
38. *New York Times,* July 10, 1974, p. 1.
39. *Newsweek,* July 22, 1974, p. 18; *Time,* July 22, 1974, pp. 15-17.
40. *New York Times,* July 10, 1974, p. 1.
41. Rowland Evans and Robert Novak, "Mr. Nixon's Supreme Court Strategy," *Washington Post,* June 12, 1974, p. A29.
42. D. Grier Stephenson, Jr., " 'The Mild Magistracy of the Law': U.S. v. Richard Nixon," *Intellect* 103 (February 1975): 292.

43. Robert Scigliano, *The Supreme Court and the Presidency* (New York: Free Press, 1971), chap. 2.
44. Lukas, *Nightmare,* p. 510.
45. *New York Times,* July 1, 1974, p. 10; *Time,* July 22, 1974, p. 16.
46. Lukas, *Nightmare,* p. 519.
47. Leon Jaworski, *The Right and the Power: The Prosecution of Watergate* (New York: Reader's Digest Press, 1976), p. 164.
48. *Washington Post,* September 9, 1974, p. A1.
49. Bob Woodward and Carl Bernstein, *The Final Days* (New York: Simon and Schuster, 1976), p. 176.
50. *New York Times* staff, *The End of a Presidency* (New York: Bantam, 1974), p. 324.
51. For a discussion of the importance of the majority-rule principle in the Court, see Thomas J. Norton, "The Supreme Court's Five to Four Decisions," *American Bar Association Journal* 9 (July 1923): 417-420; John H. Clarke, "Judicial Power to Declare Legislation Unconstitutional," *American Bar Association Journal* 9 (November 1923): 689-692; Herbert Pillen, *Majority Rule in the Supreme Court* (Washington, D.C.: Georgetown University, 1924); and Charles Warren, *The Supreme Court in United States History,* vol. 1 (Boston: Little, Brown, 1924), pp. 664-670.
52. It can be reasonably argued that the president preferred to comply with a unanimous decision (IV) than a "nondefinitive" ruling that he had been threatening to ignore (I), so (IV,I). The reversal of the ranking of the two compliance outcomes leads to essentially the same results as I shall subsequently describe, except that the equilibrium becomes (3,3) rather than (3,2), and the paradox of rational choice discussed later disappears.
53. Lukas, *Nightmare,* p. 519.
54. Ibid.
55. Interestingly, *no* outcomes in the Figure 7.5 matrix—not those associated with (4,3) or (3,2)—are in equilibrium, but this fact is not relevant to the present analysis because I have already established that this representation does not depict the game that was actually played.
56. H. R. Haldeman corroborated Jaworski's view, quoted in Section 7.7, about the consequences of a nonunanimous Court decision: "If the Supreme Court had handed down a [nonunanimous] majority decision, Nixon would have defied the Court and refused its order to turn over the tapes." H. R. Haldeman with Joseph DiMona, *The Ends of Power* (New York: Times Books, 1978), p. 310.
57. *End of a Presidency,* p. vii.
58. A good recent treatment of these situations can be found in Russell Hardin, *Collective Action* (Baltimore: Johns Hopkins University Press, 1982); see also Brian Barry and Russell Hardin, ed., *Rational Man and Irrational Society? An Introduction and Sourcebook* (Beverly Hills, Calif.: Sage, 1982).

Coalitions 8

8.1 Introduction

In almost all political games involving more than two players, the selection of coalition partners—not simply an individual strategy—assumes central importance. This is not true in the two-person games studied previously, wherein calculations of advantage were purely individualistic. Threat and deception strategies, as well as the strategy of conditional cooperation, did presume communication and coordination between the players, although not for the purpose of putting together coalitions in n-person games ($n > 2$, that is, the games relevant to the study of coalitions) to accomplish certain ends, such as winning elections. (A *coalition*, as the term will be used here, is a subset of players in a game.) Yet a large part of politics involves calculations of how *other* actors might enable a player to win a contest, obtain a preferred outcome, or enhance his or her influence in a particular situation.

A coalition's ability to implement a particular outcome is, of course, closely related to its power. By comparison, the power of an individual player is closely related to how crucial he or she is in the formation or maintenance of a coalition, as suggested by the notion of critical defection underlying the Banzhaf voting power index (Section 5.4).

This duality raises an interesting question: Is a player powerful because of his or her position in a coalition (for example, by being instrumental in its taking a particular stand), or does a coalition form because players are powerful through their being in it (for example, by their collective ability to enforce an agreement or obtain something of value)? In politics, it seems, the power of players in a game is inextricably linked with the coalitions of which they are members, and any scheme that attempts to disentangle the players from the coalitions they constitute necessarily simplifies reality.

If one is to try to understand the interrelated features of a complex system, rather than mask them as an undifferentiated whole, such simplification seems unavoidable. For analytic purposes, it appears that one may start either with the players and their preferences and ask *what coalitions will form* or with coalitions and ask *how their value will be apportioned among the players* as a function of their criticalness.

The first coalition model to be described is closely related to the spatial-competition models developed in Chapter 3 to analyze candidate strategies, but it allows for the expression of divergent interests *within* political parties. The question asked is what coalition of party interests will form to meet competition from the outside.

This question is turned around in the second model, wherein parties and candidates, as well as other players in political games, are assumed to be unitary actors. For these actors the question is not what coalition of interests will solidify internally but instead how large a coalition should be to ensure a stable winning outcome.

In the third coalition model, maximization of a player's share of the spoils will be introduced as a goal of potential coalition members. I shall derive from this model conclusions about the optimal timing of a player's actions that will help shed light on the dynamics of coalition-formation processes in which actors desire not only to win but also to benefit individually in a winning coalition. In a related model, bandwagon curves will be developed and studied. Applications of these models particularly to party and election politics in the United States, but also to international politics, will illustrate the fluidity of coalition politics—and the concomitant lack of permanent winners—when there is genuine competition in the political arena.

8.2 Political Parties as Coalitions

American political parties have a long and colorful history. Although there is general agreement that the major parties embrace a curious cast of characters, their role in the American political system is much disputed, in part because they are such heterogeneous entities.

For purposes of modeling, assume that parties contain three distinguishable sets of players: professionals, activists, and voters. The *professionals* are elected officials and party employees who have an obvious material stake in the party's survival and well-being. The *activists* are amateurs—either voters or candidates—who volunteer their services or contribute significantly to the party, especially during elections. The *voters,* who make up the great mass of the party, generally do not participate in party activities, except to vote or possibly make small contributions.

It is this mixture of groups of players, each with its own diverse interests, that makes the parties sometimes appear to be three-headed monsters—not so much because parties are terrifying creatures but rather because they are so hard to control. Often moving in different directions, parties may usefully be thought of as coalitions of players whose members somehow must reach agreement among themselves if they are to be effective political forces.

What complicates the process of reaching agreement is that the activists *tend* to take more extreme ideological positions than the professionals and ordinary voters. There are exceptions, of course, but I assume in the subsequent analysis that activists give their support because they believe in, or can gain from, the adoption of more extreme policies than the median voter would support. Not only do these policies generally give them certain psychic or material rewards, but they also usually exclude others from similar benefits. Activists tend to be purists, and they are not generally satisfied by something-for-everyone compromise solutions.

Professionals, on the other hand, are interested in the survival and well-being of their party, and they do not want to see its chances or their own future employment prospects jeopardized by the passions of the activists. Their positions generally correspond to those of the median voter, whom they do not want to alienate by acceding entirely to the wishes of the activists.

Yet, by virtue of the large contributions the activists make to the party, activist interests cannot be ignored. The election outcome would presumably be imperiled if the professionals, who are mainly interested in winning, lost the support either of the activists or of the voters.

What is the outcome of such a medley of contradictory forces? Before possible outcomes can be analyzed, the goals of candidates in elections—what they seek to maximize, given the conflicting interests of the various groups whose support they seek—must be specified.

8.3 Reconciling the Conflicting Interests within Parties

In Chapter 3 I analyzed candidate positions that were both optimal and in equilibrium in relation to those of one or more other candidates. After the nomination of one candidate by each of the major parties, the election is usually reduced to a contest between only two serious contenders. Assume that a candidate, to generate financial support (primarily from activists) and electoral support (primarily from voters) in the general election, tries to stake out positions—within certain limits—that satisfy, or at least appease, both activists and ordinary voters.

To model the candidate's decision, ignore for now the positions that the other major-party candidate may take. While the positions of a candidate's opponent will obviously determine in part his or her own positions as the campaign progresses, assume in the subsequent analysis that a party nominee's top-priority goal is to consolidate support within the ranks of his or her own party. Assume that a candidate, to satisfy this goal, cannot afford to ignore the concerns of either the activists or the voters. Without the support of the former, a candidate would lack the resources to run an effective campaign; without the support of the latter, the candidate's appeal would be severely limited even if his or her resources were not.

Consequently, assume that a party nominee seeks to maximize both resources and appeal, the former by taking positions that increase his or her attractiveness to activists and the latter by taking positions that increase his or her probability of winning among voters. Specifically, if resources (contributed by activists) are measured by the utility (U) activists derive from the candidate's positions, and if appeal (to voters) is measured by the probability (P) that these positions—given sufficient resources to make them known—will win the candidate the election, then the goal of a candidate is to take positions that maximize his or her expected utility (EU), or the product of U and P:

$$EU = U(\text{to activists})P(\text{of winning among voters})$$

EU provides a measure of the combined activist and voter support a candidate can generate from taking particular positions in the general election.

Maximization of EU implies seeking a compromise satisfactory to both the activists and the voters. Normally, this compromise will be aided by professionals who seek to reconcile the conflicting interests of the two groups; the form of this reconciliation will obviously depend on the nature of the conflicting interests that divide the activists and the voters.

8.4 Optimal Candidate Positions in a Campaign

For simplicity, assume that the campaign involves a single issue and that a candidate from the left-oriented party may take positions on this issue ranging from the left extreme (LE) to the median (Md), as shown on the horizontal axis in Figure 8.1. Assume further that the utility (measured along the vertical axis) that activists derive from the positions a candidate takes falls linearly from a high of 1 at LE to a low of 0 at Md. On the

Figure 8.1 Utility and Probability of Candidate Positions

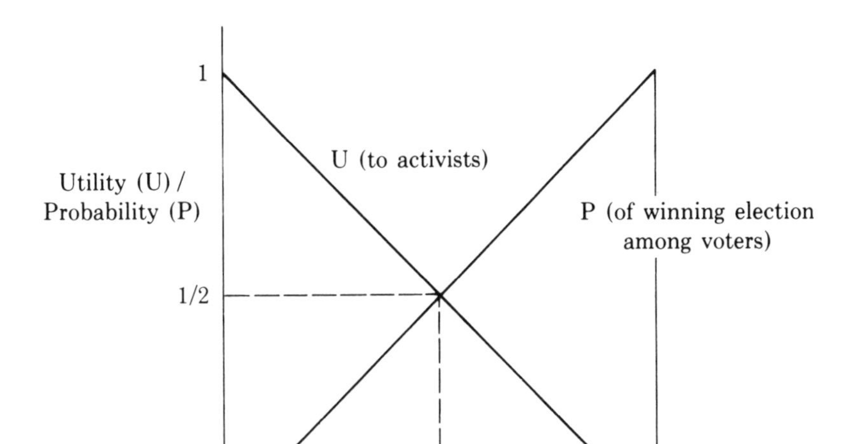

other hand, assume that the probability of winning (also measured along the vertical axis) varies in just the opposite fashion, starting from a low of 0 at LE and rising to a high of 1 at Md.[1]

I assume that the maximum probability of winning cannot be attained, however, unless adequate resources are contributed by activists to publicize the nominee's positions. Since a left-oriented activist derives 0 utility from a candidate who takes the median position, it seems reasonable to assume that no resources will be contributed to a left-oriented candidate whose position is at Md.

A candidate increases resources, but decreases the probability of winning, as he or she moves toward the left extreme. Clearly, if the candidate moves all the way left to LE, P = 0, just as U = 0 at Md. Thus, a candidate who desires to maximize EU would never choose positions at LE or Md where EU = 0.

In fact, it is possible to show that the optimal position of a candidate is at the center (C) of Figure 8.1—that is, the point on the horizontal axis midway between LE and Md where the lines representing U and P intersect. Since this point is also midway between 0 and 1 on the vertical axis, at the point 1/2,

$$EU = (1/2)(1/2) = 1/4$$

There is no other point on the horizontal axis at which a candidate can derive greater EU. Consider, for example, the point midway between C and Md, where U = 1/4 and P = 3/4. At this position,

$$EU = (1/4)(3/4) = 3/16$$

which is less than EU = 1/4 at C.

The optimality of position C in Figure 8.1 may be upset if U and P are not linear functions of a candidate's position (that is, functions that can be represented by straight lines) but instead are curves like those shown in Figure 8.2. As in Figure 8.1, the utility of a candidate's position decreases, and the probability of winning increases, as the candidate moves from LE to Md. Now, however, since U and P are not linear functions of a candidate's position along the horizontal axis, the point of intersection of the P and U curves at C on the horizontal axis may no longer be optimal.

As an illustration of this proposition, calculate EU at C and at points to the left and right of C. Clearly, at C in Figure 8.2,

$$EU = (1/3)(1/3) = 1/9 = 0.111$$

but at L (to the left of C),

$$EU = (1/2)(1/4) = 1/8 = 0.125$$

Figure 8.2 Nonlinear Utility and Probability Functions

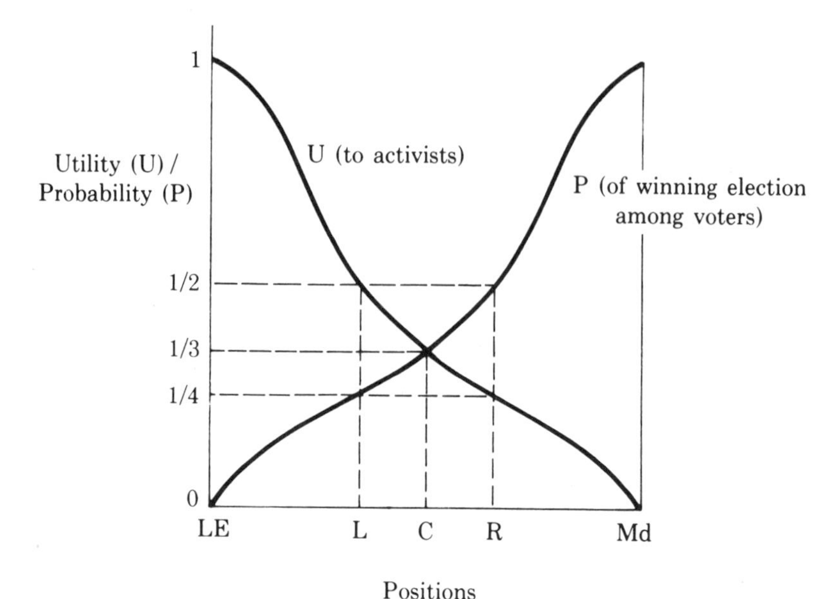

and at R (to the right of C),

$$EU = (1/4)(1/2) = 1/8 = 0.125$$

Hence, given the nonlinear utility and probability functions shown in Figure 8.2, a candidate can do better by taking a position *either* to the left or to the right of C.

The exact positions along the horizontal axis that maximize EU for a candidate depend on the shape of the U and P curves. These optimal positions can be determined from the equations that define the curves, but since there is no empirical basis for postulating particular functional relationships between candidate positions and U and P, I shall not pursue this matter further here.

The main qualitative conclusion to be derived from this analysis is that there is nothing sacrosanct about the center position C. Depending on the shape of the U and P functions, a candidate may do better—with respect to maximizing EU—by moving either toward LE or toward Md.

Whatever the shape of the U and P functions, however, if $P = 0$ at LE and $U = 0$ at Md, the positions at LE and Md will *never* be optimal since $EU = 0$ in either case. But as long as U decreases monotonically from LE to Md (that is, does not change direction by first decreasing and then increasing), and P increases monotonically from LE to Md, any points in between LE and Md may be optimal, depending on the shape of the U and P curves.

If these curves are symmetric, there may be two optimal positions, one on each side of C. Yet symmetry is not a sufficient condition for there to be more than one optimal position: the straight lines in Figure 8.1 are symmetric, but the only position along the horizontal axis where EU is maximized is at C.

What are the implications of this analysis? If activists prize extremeness, and ordinary voters prize moderation, then any position in between may be optimal for a candidate who desires to maximize some combination of resources (from activists) and electoral support (from voters). More surprising, there may be different optimal positions, one more palatable to the activists and one more palatable to the voters, as illustrated in Figure 8.2.

8.5 Optimal Positions in Actual Presidential Campaigns

So far I have shown how a simple expected-utility model might offer an explanation for the optimality of different positions in a campaign. The utility and probability functions postulated may, depending on their shape, push candidates toward an extreme position, the median position, or a center position somewhere in between.

In examining recent presidential campaigns, one can observe a variety of positions that nominees of both major parties have adopted. Barry Goldwater, the 1964 Republican nominee, and George McGovern, the 1972 Democratic nominee, provide the best examples of candidates who took relatively extreme positions in their campaigns. Both candidates had strong activist support from the right and left extremes, respectively, of their parties in the primaries, which they almost surely would have lost had they tried to move too far toward the median voter in the general election. In addition, given the moderate opposition both candidates faced from strong incumbents in the general election, neither Goldwater nor McGovern probably stood much chance of picking up many voters near the median had he tried to shift his early extremist positions very much.

If Goldwater had run against John F. Kennedy rather than Lyndon B. Johnson in 1964, however, he probably would have been a viable candidate. He could have carried all the South and West and some of the Midwest and, conceivably, might have won. Against Johnson, though, he was a loser because he and Johnson appealed in great part to the same interests; in addition, the old Kennedy voters were stuck with Johnson, which gave the incumbent a large advantage. Goldwater planned his strategy when Kennedy was alive, and he could not jettison it after Kennedy was assassinated.

By comparison, McGovern's early extremist positions were no match from the beginning for Richard Nixon's middle-of-the-road positions. When, in desperation, McGovern attempted to moderate some of his early positions, he was accused of being wishy-washy and probably suffered a net loss in electoral and financial support (see Section 3.7).

In general, if the utility for activists falls off rapidly and the probability of winning increases only slowly as a candidate moves toward a median, his or her optimal position will be near the extreme. Such a position, compared with a position near the median, gains the candidate more in resources than he or she loses in probability of winning. With this trade-off in mind, both Goldwater and McGovern seem to have acted rationally with respect to the maximization of EU, though McGovern was apparently more willing to sacrifice activist support to increase his chances of winning.

The incumbent presidents Goldwater and McGovern faced, Johnson and Nixon, had more moderate activist supporters who were less disaffected by middle-of-the-road politics. Not only could these incumbents afford to move toward the median voter and still count on significant activist support, but also, because of the extreme positions of their opponents, they could probably rapidly increase the number of their moderate supporters with such a strategy.

However, as Coleman has pointed out, if an incumbent already has greater a priori strength than his opponent—and his opponent magnifies the discrepancy in strength by adopting an extremist position—the incumbent will not significantly improve his (already high) probability of winning by moving farther away from the other extremist position and toward the median.[2] Against such an opponent, therefore, an incumbent with a large built-in advantage from the start has little incentive to move toward him. Thus, extremist positions, especially when there is such an a priori difference in electoral strength, will tend to reinforce each other: both candidates will be motivated to adopt relatively extreme positions, because movement by one candidate toward the other decreases his activist support more than it increases his probability of winning.

Although this conclusion follows logically from the EU model, it seems to have little empirical support. The Goldwater-Johnson and McGovern-Nixon races did not produce extremists on both sides, only on one. In fact, if one candidate's position diverges sharply from the median, as did those of Goldwater and McGovern, his or her opponent tends to move toward that position rather than stay the same or move in the opposite direction.

This behavior is explained quite well by the earlier spatial models (see, in particular, Section 3.3), but it is difficult to derive it from the goal of maximization of EU in which P is one factor. After all, if P is already high for a strong incumbent running against an opponent who adopts an extremist position, why should the incumbent move toward his or her opponent if this movement has little effect on P and may lower U at the same time?

The answer seems to lie in the fact that some candidates seem to be as interested in the absolute size of their majorities (or pluralities) as in winning. That is, they desire large majorities/pluralities at least as much as victory itself. If such is the case, then movement toward an extremist opponent can be explained by the fact that this movement steadily increases a strong candidate's vote total even if it does not significantly alter his or her (already high) probability of winning.

Both Johnson's and Nixon's campaigns strongly indicate that, even with victory virtually assured months before the election, they wanted more than victory: they desired to pile up huge majorities by whatever means they had at their disposal—including misrepresentation of their positions and those of their opponents. Indeed, both incumbents succeeded in crushing their opponents in their respective elections, although both were later driven from office by a welter of forces that will be analyzed later (see also Sections 7.7 and 7.8 on the more proximate causes of Nixon's departure).

If the goals presidential candidates seek to maximize preclude *both* candidates from diverging from the median—and may encourage convergence, as in the 1960, 1968, and 1976 elections—then it is unlikely that one of the major parties can be written off the national political scene for very long. Indeed, in recent presidential elections, there has been a steady alternation of ins and outs: no party since World War II has held office for more than two consecutive terms, whereas earlier one party or the other sometimes held the presidency for a generation or more.

The parties alternated again in the 1980 presidential election, but with one significant difference: an elected incumbent president lost his reelection bid for the first time since 1932. The 1980 election is also distinguished in that Ronald Reagan was nominally a conservative, not a moderate, but I think this is a misnomer: not only was he strongly favored by his own party (Republican) in the nomination race but he was also more acceptable to a majority of the electorate than Jimmy Carter, who was saddled with a failing economy and the American hostage situation in Iran. In the end, Reagan was able to marshal enthusiastic conservative support and also capture many moderates because the major alternative was unacceptable (John Anderson, the independent candidate in this race, got less than 7 percent of the popular vote).

Then, with a vastly improved economy in 1984, Reagan triumphed again. Without an incumbent in 1988, the Republican party should face stiffer opposition, and the parties may again alternate.

8.6 The Size Principle

So far in this analysis I have assumed that political candidates seek to maximize both activist support and their probability of winning. I have suggested that the latter objective could be modified to reflect the desire—apparently true of certain candidates—to maximize their vote totals, which may only marginally affect their probability of winning beyond a certain point.

In certain situations political actors seem to have more straightforward desires. In his model of political coalitions, Riker divorces winning from other objectives and postulates it to be *the* goal of "rational political man":

> What the rational political man wants, I believe, is to win, a much more specific and specifiable motive than the desire for power.... The man who wants to win also wants to make other people do things they would not otherwise do, he wants to exploit each situation to his advantage, and he wants to succeed in a given situation.[3]

Because winning helps an actor achieve these things in elections, wars, and other arenas of political conflict, it is not surprising that he or she places a high value on that goal. Moreover, as a concept for realizing a whole panoply of ends, the notion of winning offers the political theorist a generalized goal for political actors that conveniently subsumes particular goals that are often difficult to know or specify in actual situations.

Any statement about a postulated goal is necessarily based on the *perceptions* of actors. Ordinarily, this causes no problem in political situations like elections, where the method of counting votes and the decision rule for selecting a winner are known and accepted by the contestants—at least in situations where information is assumed to be complete. But it should be noted that although elections usually distinguish unambiguously the winner from the losers, the winner may not be perceived as the true victor, as when a candidate does better than expected in a presidential primary and is declared the nominal winner, in spite of receiving fewer votes than an opponent.[4]

Besides positing the goal of winning, Riker makes several assumptions in his game-theoretic model (the mathematical details of which will be ignored here):

1. *Rationality.* Players are rational, which means that they will choose the alternative that leads to their most-preferred outcome—namely, winning.[5] Riker, however, argues not that all political actors are rational with respect to this goal but rather that the winner-loser mentality pervades, and conditions the behavior of, participants in such situations as elections and total wars.
2. *Zero-sum.* Decisions have a winner-take-all character—what one coalition wins the other coalition loses—so the sum of payoffs to all players is zero. In other words, the model embraces only situations of pure and unrelieved conflict where all value accrues to the winner; cooperation among participants that redounds to the mutual benefit of all is excluded.
3. *Complete and perfect information.* Players are fully informed about the state of affairs at the beginning of the game and about the moves of all other players throughout the game.
4. *Allowance for side payments.* Players can communicate with each other and bargain about the distribution of payoffs in a winning coalition, whose value is divided among its members. (*Side payments* are simply individual payments players transfer to each other in dividing up the value.)
5. *Positive value.* Only winning coalitions have positive value.
6. *Positive payoffs.* All members in a winning coalition receive positive payoffs. This assumption, of course, provides an incentive for players to join a winning coalition.

7. *Control over membership.* Members of a winning coalition have the ability to admit or eject members from it.

Assumptions 5, 6, and 7—besides the assumption that the goal of players is to form winning coalitions—are what Riker calls "sociological assumptions," as distinguished from the four "mathematical assumptions" (1, 2, 3, and 4) standard in n-person game theory. The sociological assumptions specify more precisely the goal of winning and thereby enable Riker to derive the size of winning coalitions that is optimal and therefore likely to occur.[6]

Given these assumptions, Riker shows that there are no circumstances wherein an incentive exists for coalitions of greater than minimal winning size to form. On the other hand, the fact that there is a positive value associated with winning coalitions (assumption 5), each of whose members receive positive payoffs from winning (assumption 6), is a sufficient incentive for such coalitions to form. The incentive for winning coalitions to form, but not to be of greater than minimal winning size, means that the realization of the goal of winning takes form in the creation only of minimal winning coalitions; Riker calls this conclusion the *size principle.*

His reasoning is as follows. Given that only winning coalitions have positive value (assumption 5), the zero-sum assumption (assumption 2) implies that losing coalitions must have complementary negative value. Since such a coalition has no positive value to distribute among its members, it would form only as a pretender to eventual winning status. Indeed, the possibility that a losing coalition could eventually become winning provides a strong incentive for a winning coalition to pare off superfluous members (permitted by assumption 7) before they and other disaffected members to whom it cannot offer sufficient payoffs defect. If the excess members are not ejected, the winning coalition becomes vulnerable to offers from a losing coalition that could promise enough defectors greater rewards in a (prospective) minimal winning coalition so as actually to constitute such a coalition. Because the minimal winning coalition has complete and perfect information (by assumption 3), its members know exactly when they have enough members in the coalition to win.

Several things should be noted about Riker's derivation of the size principle. First, winning is not itself an explicit goal: rationality (assumption 1) motivates players to obtain the benefits of being in a winning coalition, as stipulated in assumptions 5 and 6. Second, the size principle is a statement about an outcome—the size of winning coalitions—and not about the process of coalition formation, which more will be said about in the remainder of this chapter. Finally, since any win is the sole determinant of value (assumption 5), the ejection of superfluous members from a winning coalition means that the same total amount of value can be divided among the fewer members of a minimal winning coalition.

The idea that the smaller the size of a winning coalition, the more each of its members individually profits and would therefore be expected to work to reduce its size would seem not to require the formalism of a mathematical model to grasp. Yet this commonsensical explanation of the size principle does not offer limiting conditions on the veracity of the size principle, whereas assumptions of a model do.

As an explanation of why the size principle should hold, however, Riker's argument has logical force only. To connect Riker's model with reality, consider how its assumptions can be interpreted as conditions that limit the operation of the size principle when its theoretical concepts are operationally defined and it is posited as an empirical law. Riker offers this translation of the size principle: "In social situations similar to n-person games with side-payments, participants create coalitions just as large as they believe will ensure winning and no larger." [7]

The introduction of the beliefs or perceptions of individuals about a subjectively estimated minimum simply acknowledges the real-world fact that players do not have complete and perfect information about the environment in which they act and about the actions of other players. Consequently, players are inclined to form coalitions that are larger than minimal winning size as a cushion against uncertainty, and oversized winning coalitions thus become a quite rational response in an uncertain world.

In this manner, complete and perfect information serves as a limiting condition on the truth of the size principle: in its absence, coalitions will *not* tend toward minimal winning size. Similarly, the other assumptions of the model also restrict the applicability of the size principle in the real world, but Riker singles out the effect of information for special attention probably because it is the concept most easily interpreted, if not operationalized, of those embodied in the assumptions of his game-theoretic model.

The enlargement of coalitions to more than minimal winning size due to the effects of incomplete or imperfect information is what Riker calls the *information effect*. Since this effect is endemic in an uncertain world, there naturally exist many examples of nonminimal winning coalitions. For Riker these examples represent situations in which coalition leaders miscalculated the capabilities, or misread the intentions, of opponents—not because they were irrational but rather because they lacked accurate and reliable information on which to act.

8.7 Applications of the Size Principle to Politics

However oversized coalitions come into existence, the prediction of the size principle is that they will be relatively short-lived. In seeking out evidence that supports the principle, Riker examined all instances of "overwhelming majorities" in the modern European state system. He found three such cases, all being the products of total war in which one coalition of states became dominant in the system upon defeating an opposing coalition. These examples of overwhelming majorities in world politics—the Concert of Europe allies (England, Austria, Prussia, and Russia) after the Napoleonic wars, and the Allied powers after World War I and after World War II—arose in situations that seem to approximate the assumptions of the model (except for complete and perfect information). During the wars the governments on each side were intent on destroying the governments on the other side, which suggests that the wars were basically zero-zum in character; nations for the most part sought allies who would enhance their capabilities of winning; and side payments reckoned in promises and threats, as well as material goods, were exchanged.

The conclusion of the total wars in all of these instances resulted in a large coalition of winners, which was essentially worthless since there was no significant further value to extract from the losers. Accordingly, each of these coalitions was almost immediately plagued by internal strife, with the Congress of Vienna splitting into two camps (Austria and England versus Prussia and Russia) after the Napoleonic wars; England, France, and the United States dividing over the future role of Germany after World War I; and the Soviet Union and the United States fighting for allies and hegemony after World War II. The hopes for permanent peace enshrined first in the League of Nations and now in the United Nations founder on the size principle precisely because of the undiscriminating inclusiveness of these international organizations. Grand coalitions that comprise all the players have zero value in zero-sum games.

Riker's second source of evidence on the dissolution of overwhelming majorities is from American presidential politics. As in total wars, victory in elections is indivisible; hence, Riker argues, elections can properly be modeled as zero-sum games, where the players are individuals and groups who unite behind the banners of political parties. Looking at all instances in which one party effectively demolished another party in a presidential election and the losing party virtually disappeared from the national scene, he found that the period of one-party dominance following such elections is soon undercut by party leaders who force out certain elements of the party in the course of policy and personality disputes. This leads to the dominant party's shrinkage; then, usually after some miscalculation, it is displaced by another party.

The three instances of overwhelming majorities in American politics analyzed in Riker's *Theory of Political Coalitions* are the Republican party after the 1816 election that destroyed the Federalist party, the Democratic party after the 1852 election that destroyed the Whig party, and the Republican party after the 1868 election that signaled the temporary demise of the Democratic party, at least outside the South. Again he shows that oversized coalitions did not endure.

Riker and Ordershook have elsewhere analyzed the aftermath of the 1964 election, in which Johnson beat Goldwater by a 16-million-vote margin, in terms of the size principle. Arguing that Johnson dissipated his overwhelming majority by progressively alienating many southerners with his strong pro-civil rights stance and then liberals with his escalation of the Vietnam war, Riker and Ordeshook contend that by 1968 Johnson probably did not have the support of even a minimal majority of the electorate and hence chose not to run for a second term.[8]

A similar fate befell Richard Nixon after his landslide victory in 1972. Unable to extricate himself from the albatross of Watergate, he was forced to resign 18 months into his second term. It seems no accident that whatever the time and circumstance, every U.S. president or political party that has emerged with an overwhelming majority has always faced a mounting tide of opposition to which the president or political party has succumbed in the end. In zero-sum politics, it appears, instability occasioned by the miscalculations of leaders is unavoidable in a world enshrouded by uncertainty.

Since the publication of Riker's *Theory of Political Coalitions,* the size principle has stimulated many efforts aimed at testing its truth in a variety of empirical settings.[9] It has also been the target of much criticism, with one of the most frequently voiced complaints being its lack of specificity in describing how long an oversized coalition may be expected to persist. To respond that this will depend on the incompleteness or imperfectness of information is not helpful when no time-dependent functional relationship between this variable and the size principle is specified. Will the excess majority of an oversized coalition be dissipated in weeks, years, or decades?

Riker himself admitted that the model is "quite vague" on this point and that specifying a pattern in the growth of coalitions "may be regarded as the main task of a dynamic theory of coalitions." [10] To probe more deeply into the dynamic nature of transitions from stage to stage, goals different from winning to which rational actors might aspire must be postulated and their implications developed.

8.8 An Alternative Goal: Maximizing One's Share of Spoils

Although the study of coalition behavior has been vigorously pursued in recent years, little attention has been devoted to the construction of theoretical models of coalition-formation processes that occur over time. Rather, studies of coalition behavior have focused more on *static outcomes* than on the *dynamic processes* that produced them *prior* to the point at which one coalition has gone on to win. Even in the rule-bound setting of voting bodies, to which the subsequent analysis will be confined, the development of dynamic models has only recently been initiated.

This gap in the literature is due in considerable part to the failure of theorists to postulate goals for rational actors that raise questions about the *timing* of their actions—and therefore about political processes that occur over time. Before postulating such a goal, I shall first outline a verbal model of coalition-formation processes in voting bodies.

The analysis will be restricted to the study of coalition-formation processes involving as active opponents only two *protocoalitions*—coalitions not of sufficient size to win—that vie for the support of uncommitted members in a voting body so that they can become winning coalitions (according to some decision rule). As before, it is useful to distinguish all winning coalitions from the subset of winning coalitions that are minimal winning—coalitions in which the subtraction of a single member reduces them to (nonwinning) protocoalitions.

My main interest is in studying the formation of winning coalitions including one or the other of the two protocoalitions, but not both. I assume that the two protocoalitions are totally at odds with each other and seek victory only through securing the commitment of uncommitted members, not through switching members' commitments from one protocoalition to another. Members of the two protocoalitions are thus precluded from combining to form a winning coalition.

As an illustration of this model, consider the presidential election of 1824, which is analyzed in detail by Riker.[11] None of the four major candidates was able to win a majority in the electoral college, so the election was thrown into the House of Representatives in 1825, where each state had a single vote that went to the candidate favored by a plurality of its representatives. The standing of the four candidates before voting began was

John Quincy Adams	10 votes (states)
Andrew Jackson	7 votes (states)
William Crawford	4 votes (states)
Henry Clay	3 votes (states)

a situation that would have forced the elimination of Clay under the Twelfth Amendment, which allows for only three runoff candidates.

Realizing that the votes Clay controlled were open for bidding, Crawford's managers, it seems fair to suppose, might also have considered selling their votes since their candidate ranked well below the two front-runners.[12] If the two leading candidates, Adams and Jackson, represent the two protocoalitions in the model I have described and the "uncommitted" votes are the blocs controlled by Clay and Crawford, it is simple to determine how each of the two leading candidates could have won (the decision rule is 13 out of 24 states):

Adams, with the support of (1) Clay, (2) Crawford, or (3) both Clay and Crawford
Jackson, with the support of (4) both Clay and Crawford

Without any additional information about coalitions that might form in this situation, it is reasonable to assume that each of these four outcomes is equally likely (this assumption will later be modified in light of the size principle). Hence, the complementary probabilities (Ps) that each of the two leading candidates (that is, protocoalitions) would become winning can be calculated:

$$P(\text{Adams}) = 3/4 = 0.75; \; P(\text{Jackson}) = 1/4 = 0.25$$

In fact, Clay threw his support to Adams as a result of the "corrupt bargain"—a promise from Adams to make him secretary of state—making Adams the winner (with probability equal, in effect, to 1.00). If Clay had instead supported Jackson, then Jackson and Adams would have tied with 10 votes each. If, at this hypothetical juncture, Crawford had been equally likely to have given his support to either leading candidate, then each would have had a probability equal to 0.50 of going on to win.

The two choices open to Clay—support Adams or support Jackson—are pictured in Figure 8.3, with the probabilities that each of the main contenders, Adams and Jackson, would win before and after Clay's commitment to one of them.[13] Along each branch in Figure 8.3 are shown the *probabilistic contributions* that Clay's support makes to raising the probability of winning of Adams (from 0.75 to 1.00, or by an increment of 0.25) and of Jackson (from 0.25 to 0.50, or by an increment of 0.25). The fact that these probabilistic contributions are both equal to 0.25 means that no advantage would accrue to Clay from supporting one candidate over the other if his goal were to maximize his probabilistic contribution to a contender.

The goal I postulate, however, is that uncommitted actors seek to maximize their *share of spoils* (SS), which is defined as the probabilistic

Figure 8.3 Probabilities of Becoming Winning, and Probabilistic Contributions, in 1824-1825 Presidential Election

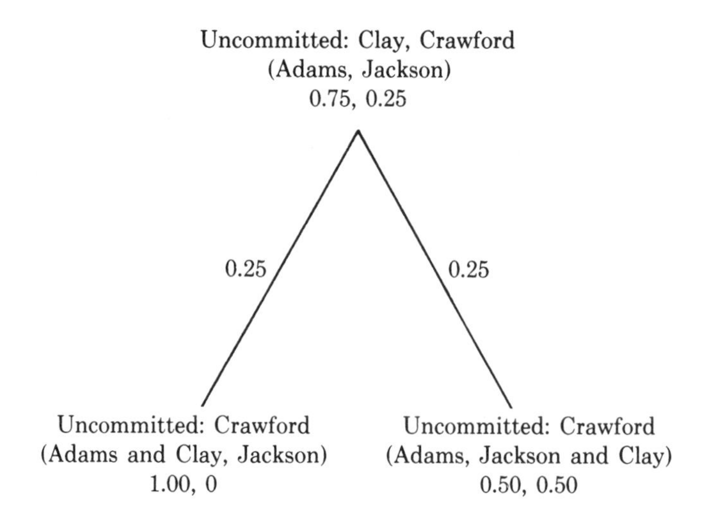

contribution (PC) an uncommitted actor makes to a protocoalition multiplied by the probability (P) that, with this contribution, the protocoalition will go on to win. Thus

$$\begin{aligned} SS(\text{Clay} \rightarrow \text{Adams}) &= PC(\text{Clay} \rightarrow \text{Adams})P(\text{Clay} \rightarrow \text{Adams}) \\ &= (0.25)(1.00) = 0.25 \\ SS(\text{Clay} \rightarrow \text{Jackson}) &= PC(\text{Clay} \rightarrow \text{Jackson})P(\text{Clay} \rightarrow \text{Jackson}) \\ &= (0.25)(0.50) = 0.125 \end{aligned}$$

where the actor before the arrow indicates the uncommitted "giver" and the actor after the arrow the protocoalition "receiver." Assuming that Clay's goal was to maximize his share of spoils, then it was rational by this calculation for him to support Adams rather than Jackson (as he in fact did), since the former action would yield twice as much in spoils as the latter.

The probabilities computed above reflect whether the two leading candidates would become *any winning* coalitions—that is, coalitions with *at least* a simple majority of members—before and after the commitment of Clay. If one restricts coalitions to those that are *minimal winning* with *exactly* a simple majority of members, then consonant with the size principle there is only one way that Adams can win with exactly 13 of the 24 votes (that is, by gaining the support of Clay), but no way Jackson can win with this bare majority.[14] Since the Adams-Clay alliance is the only possible minimal winning coalition, it will become minimal winning with probability equal to 1.00 *before* Clay's commitment, given that only minimal winning coalitions form. *After* Clay's commitment, the coalition will be winning (with probability equal, in effect, to 1.00), so Clay's probabilistic contribution, and share of spoils that he can expect to receive from Adams, is 0.

This may seem a rather artificial result, especially in light of the fact that Clay became secretary of state under Adams and was importuned by all parties for support before the House vote.[15] Apparently, not everybody was aware of the size principle, for if they had been all except Adams's supporters should have given up hope!

Despite its artificiality, however, this result points up one nonobvious fact about coalition formation: to the extent that an uncommitted actor is constrained in the range of choices open to him or her, his or her influence will be diminished. Indeed, in the extreme case where the actor is a captive of one protocoalition, his or her probabilistic contribution and share of spoils are effectively 0 since he or she is not truly uncommitted. This was the case in the 1824-1825 presidential election, for if Clay adhered to the size principle, he could support only Adams.

In other words, the size principle, like friendship, ideology, and other constraints on free association, may cut down an actor's range of choices and force his or her hand before all alternatives but one have been foreclosed. In fact, one month before the actual balloting in the House, Clay confided to intimates that he would support Adams,[16] a fact that became public knowledge two weeks before the election.[17] But there was still great uncertainty about what the outcome would be because the New York delegation (counted earlier in Adams's total) was deadlocked up to the day of voting. The election eventually turned on the vote of one of the 34 representatives from New York, who on the day of voting broke the deadlock and gave New York to Adams.[18]

How do these historical details bear on models of coalition formation? First, I have shown that an alternative goal—maximization of share of spoils—also retrodicts (predicts a past state of affairs) that Clay would support Adams over Jackson. In the example considered, this goal is perhaps more realistic than the goal of winning because it directly incorporates the notion of private and divisible benefits that uncommitted actors may realize in addition to the public and indivisible benefits of winning that all members of the winning coalition share. Second, I have shown some curious implications of the share-of-spoils calculation when the size principle is invoked. If all but minimal winning coalitions are disallowed, the share-of-spoils calculation suggests that Clay's votes were not essential to Adams, for the size principle precludes their going to anybody else. Manifestly, the situation was not viewed this way by the participants, although Clay's votes were committed before the day of the election and it fell to somebody else to make the critical choice on that day. This leads one to ask: What makes a player's choices critical?

Choices of players are *critical,* it seems, if (and only if) they are preceded by prior choices of other players that make them decisive in the determination of outcomes. For this reason, it is useful to view coalition-formation processes as *sequences of moves* (compare Section 6.6) on the part of players. In most real coalition-building situations, however, it is no easy task to abstract any set of general contingencies.[19] Even in a particular situation, the possible alignments and the ways in which they can form are manifold. What can be offered, however, is a prescription of how protocoalitions should grow *at all stages* in the coalition-formation process to maximize their attractiveness to uncommitted actors.

Intuitively, it seems clear that uncommitted actors who wish to maximize their share of spoils in the previously described model would be interested in joining the protocoalition with the greater probability of winning. Otherwise, their probabilistic contributions would be cut by more than half in the share-of-spoils calculation. On the other hand, if the probability that one of the two protocoalitions would become winning were overwhelming (that is, close to 1.00), the probabilistic contribution that an uncommitted actor would make by joining it would be necessarily small, even though the contribution itself would not be heavily discounted in the share-of-spoils calculation.

It turns out that if there are two protocoalitions, one of which grows faster than the other, an uncommitted actor should join the one he or she thinks is growing faster—as measured by its larger size at any stage—when its probability of becoming *minimal winning* is 2/3.[20] That is, the actor should wait until the probability of one protocoalition's becoming minimal winning is exactly twice that of the other's becoming winning before making a commitment to what seems to be the faster-growing protocoalition. This advice is applicable whether this point (or points) is reached early or late in the coalition-formation process.

From the viewpoint of the coalition leaders, an awareness of this strategy provides no guarantee of victory. For example, a protocoalition may simply not have the resources to pull far enough ahead of another protocoalition to be perceived as the two-to-one favorite. If it *is* able to accomplish this feat, however, the "2/3's rule" prescribes that it should not attempt to increase these odds in its favor still further, for then it will become less attractive to uncommitted actors who desire to maximize their share of spoils. One would expect, therefore, that a protocoalition will not only maximize the commitments it receives from uncommitted actors when it enjoys a two-to-one probabilistic advantage over the other protocoalition but also keep the race exciting if it does not pull ahead much further.

Of course, an opposition protocoalition would be attempting to achieve the same kind of effect, which means that neither protocoalition has a surefire winning strategy unless additional assumptions are made about the total resources possessed by each side, the manner in which they are allocated, and so on. By the same token, uncommitted actors cannot anticipate whether and when optimal commitment times will occur in the absence of additional assumptions. The 2/3's rule as such says only that there exist optimal commitment times, but it says nothing about instrumental strategies for realizing them. Still, this rule offers a precise quantitative statement of when bandwagons would be expected to develop and provides, in addition, a rationale for their existence based on the assumptions of the model from which it is derived (which I shall not develop rigorously here).

8.9 The Bandwagon Curve

Philip D. Straffin, Jr., has argued that the assumptions underlying the 2/3's rule are quite restrictive, and he has proposed an alternative model to analyze coalition formation in large voting bodies.[21] Straffin's model is based on the application of a modified Shapley-Shubik power index,[22] which is similar to the Banzhaf index of voting power (discussed in Section 5.4), to oceanic games. *Oceanic games* are games in which there are two—or possibly more—"large players" (major candidates) who control the committed voters and an "ocean"of uncommitted voters in an infinite-size voting body. As an illustration of the meaning of these curves in the finite case, consider a voter in a five-member voting body in which two members are already committed to candidate X and one member to candidate Y. If a majority of three votes is needed to win, is it in the interest of one of the two uncommitted members to join X, to join Y, or to stay uncommitted?

Consider all possible orders of X (two committed members), Y (one committed member), and the two uncommitted members (each designated by "1"):

*X (Y) 1 1	*Y 1 (X) 1	1 1 (X) Y
X (1) Y 1	Y 1 (1) X	1 Y (1) X
X (1) 1 Y	1 (X) 1 Y	*1 Y (X) 1
*Y (X) 1 1	1 (X) Y 1	1 1 (Y) X

In each ordering, the actor who casts the third vote, or *pivotal vote,* is shown in parentheses. This member is called the *pivot.* Following Brams and Riker,[23] Straffin excludes as illegal the four orderings (marked by an asterisk) in which *both* X and Y are members of the winning coalition—by being pivotal or to the left of the pivot—under the assumption that uncommitted members will form a winning coalition only with either X or Y but not both. Of the eight remaining orderings, X is pivotal in three, Y in one, and the two uncommitted members in four, or two each.

Define the *pivotal power* of a member to be the fraction of (legal) orderings in which he or she is pivotal. Thus, the pivotal power of the uncommitted members is 2/8 = 1/4 = 0.250, and the pivotal power of X is 3/8 = 0.375. On the other hand, the pivotal power of Y is 1/8 = 0.125.

If an uncommitted member joins X, he or she makes X winning with three members, giving X (now with three members), pivotal power of 1.000. This makes the uncommitted member's *incremental contribution* to X

$$1.000 - 0.375 = 0.625$$

(If the uncommitted member were to join Y instead, his or her incremental contribution would be 0.250 − 0.125 = 0.125 since after the member joins Y, X and Y each have two members and pivotal power of 0.250.) This means that an uncommitted member does best by joining X, for an incremental contribution of 0.625, which is more than he or she contributes by joining Y (0.125) and more power than he or she has by remaining uncommitted (0.250).

Applying this reasoning in the infinite case, Straffin is able to construct "bandwagon curves" for the major candidates, X and Y, from his model (see Figure 8.4). These curves divide the region in Figure 8.4 into subregions in which uncommitted voters should commit to X, commit to Y, or stay uncommitted. Thus, an uncommitted voter should stay uncommitted—not join X or Y—if the percentage of voters committed to X and Y is a point that falls in the subregion between the bandwagon curves of X and Y in Figure 8.4. Otherwise, the uncommitted voter should join X and Y, respectively, depending on whether (1) X is the larger protocoaliton—has more committed voters—and the situation is defined by a point in the subregion below the bandwagon curve for X, or (2) Y is the larger protocoalition and the situation is defined by a point in the subregion above the bandwagon curve for Y.

Thus, an uncommitted voter should *always* join the larger protocoalition if its lead is "sufficiently great." *How great* is given precisely by the curves, but for rough guidance Straffin provides the following rules of thumb: the advantage that the larger protocoalition needs to start a bandwagon in its favor is:[24]

4:1 if 10 percent of the members are committed
3:1 if 20 percent of the members are committed
2:1 if 40 percent of the members are committed
3:2 if 60 percent of the members are committed

Figure 8.4 Bandwagon Curves, with Week-by-Week Delegate Commitments to Ford (X) and Reagan (Y) in 1976 Republican Party Race

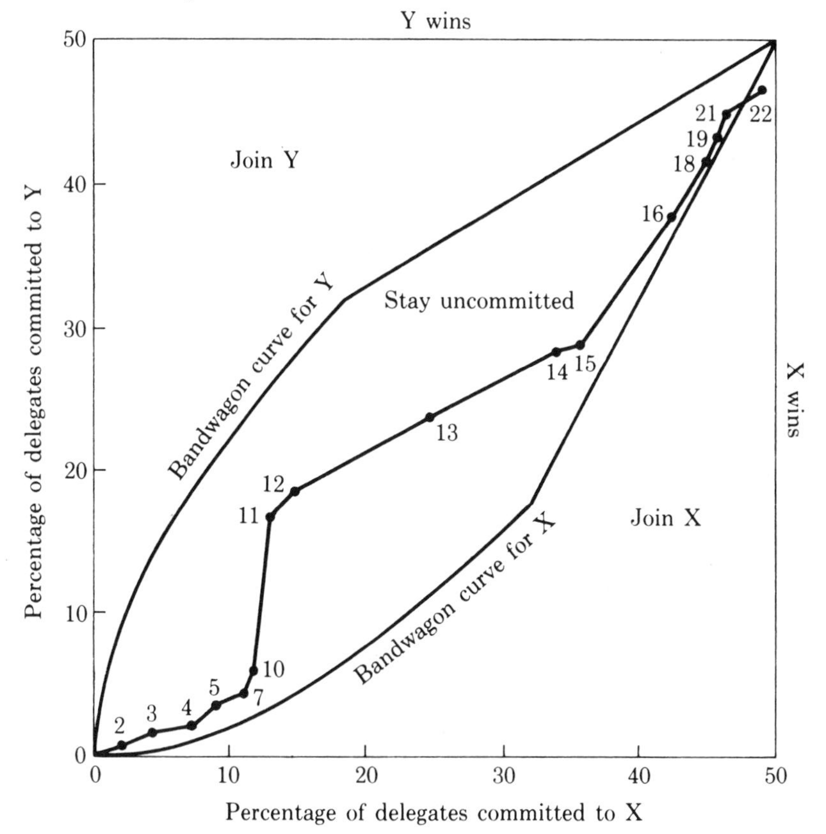

Source: Philip D. Straffin, Jr., "The Bandwagon Curve," *American Journal of Political Science* 21 (November 1977): 695-709. Copyright © 1977 by University of Texas Press. Reprinted by permission. Straffin based the week-to-week totals on reports in *Congressional Quarterly Weekly Report* and the *New York Times,* February through July 1976.

The acid test of a model, of course, is how well it works, and Straffin's model worked remarkably well in predicting the outcome of the race for the Republican party presidential nomination between Gerald Ford and Ronald Reagan in 1976. As shown in Figure 8.4, Straffin plotted the delegate totals of candidates Ford (X) and Reagan (Y) on a week-to-week basis.

Beginning with the New Hampshire primary in week 1 and continuing for the next 10 weeks, Ford was ahead but still within his bandwagon curve. Then, after the Texas and Indiana primaries in weeks 11 and 12,

Reagan was ahead—but by not nearly enough to shoot past his bandwagon curve into the bandwagon region—and his momentum petered out. Slowly, once again, Ford began to gain strength, and in week 22—specifically, on July 17 when the Connecticut delegation gave Ford 35 votes—Ford moved past his bandwagon curve into the bandwagon region. Common perceptions soon matched the mathematics of the bandwagon curve:

> On July 19, the *Washington Post* headlined "Privately, Some Aides See Reagan at End of Trail," and the story was picked up across the country. On the same day the *Los Angeles Times* reported a change in tactics by the Ford delegate hunters: "The message Ford people are now sending to uncommitted delegates is simple and direct: you had better get aboard—the train is about to leave without you." [25]

Despite Reagan's best efforts first to dispute the figures and then to muddy the waters with his surprise announcement of Richard Schweiker, a liberal Republican senator from Pennsylvania, as his choice for running mate, Ford went on to win the nomination on the first ballot of the Republican party convention on August 18. Straffin makes the following trenchant observation about the effects of the Schweiker announcement:

> The bandwagon curve analysis suggests that the Schweiker announcement was valuable not as pressure on Gerald Ford to make a similar announcement, or as a bridge to liberal Republicans, or as an attempt to gain Pennsylvania delegates, but as a successful move to divert attention from bandwagon-starting delegate counts. It also helped make those counts seem less reliable by centering speculation on how the Schweiker announcement would alter delegate commitments. The bandwagon curve provides a framework in which to view the remarkable events of July and August 1976.[26]

To conclude this analysis, it seems fair to say that goals like winning, which apply to a coalition as a whole, tend to shut off inquiry into the dynamic analysis of what offers protocoalitions are likely to make, and uncommitted actors to accept, at particular times in the formation and disintegration of winning coalitions. Simply to say that uncommitted actors seek to obtain the benefits of winning does not by itself entail the choice of a particular strategy on their part. By contrast, postulating the goal of not just obtaining the benefits of winning but also maximizing one's portion of these benefits—as measured by one's share of spoils or pivotal power—encourages the analyst to pay attention to the optimal timing of strategic choices (for example, when to join a protocoalition), and hence the unfolding of events (such as bandwagons) over time.

8.10 Conclusions

In this chapter a variety of coalition models were developed and applied to the study of American and international politics. In the case of each model, an effort was made to determine the optimal strategies that players would follow to maximize some postulated goal and to compare these with the actual behavior of parties and candidates in presidential elections and states in the international system. Mostly historical evidence was assayed in an attempt not so much to adduce anecdotal support for a model but rather to illustrate each model's broad applicability to a range of different cases.

The historical evidence considered in this chapter strongly suggests that winning alliances in total wars, as well as successful coalitions in American politics, are none too permanent and may be indeed quite fragile. The models indicate several reasons why this is the case, including

1. The influence exerted by activists who favor extremist candidates, which has led in the recent past to stunning victories and defeats for both political parties
2. Incomplete and imperfect information about what strategies will produce minimal winning coalitions, which has encouraged miscalculations resulting in oversized coalitions that eventually succumbed to the size principle
3. The unattractiveness to uncommitted actors of protocoalitions with overwhelming odds in their favor (based on the share-of-spoils calculation that says a two-to-one probabilistic advantage is optimal)

To be sure, the pivotal-power calculation *always* favors joining the front-runner over the second-leading candidate, but not necessarily over staying uncommitted if the front-runner's lead is not too large. Thus, even this model says that there is a basic fragility about a leader's position, which small shifts may suddenly upset.

For these and probably other reasons, there seems little danger that winners will become permanently institutionalized in American politics—given, of course, that the basic rules of the game allowing for the formation and disintegration of coalitions do not change. Hence, despite the occasional humiliation of presidential candidates (and, by implication, their parties) at the polls, the competitive two-party system—and the concomitant policy swings that alternating parties engender—do not seem about to disappear from the national scene.[27] International alliances are also likely to fluctuate somewhat, but barring a catastrophic nuclear war that makes everybody a loser, the type of grand coalitions that emerged from total wars in the past are probably an anachronism in the present international political arena.

NOTES

1. If a candidate's opponent also adopts a position at Md, then the candidate's P at Md will be 0.5 instead of 1.0, assuming the activist support (resources) of both candidates are the same at Md. Although the actual value of a candidate's P at Md—dependent on his or her opponent's behavior—does not affect the maximization of EU (defined in the text), it may affect strategy choices in a manner to be discussed later.
2. James S. Coleman, "Communications," *American Political Science Review* 67 (June 1973): 567-569.
3. William H. Riker, *The Theory of Political Coalitions* (New Haven, Conn.: Yale University Press, 1962), p. 22.
4. The Democratic party presidential primary in New Hampshire in 1968 is a case in point. Incumbent president Johnson captured 50 percent of the vote to Eugene McCarthy's 42 percent, with the remaining 8 percent split among minor candidates. Despite the facts that Johnson's candidacy was unannounced (his name did not even appear on the ballot) and he never set foot in New Hampshire, a majority of voters took the trouble to write in his name. Nonetheless, McCarthy was hailed as the victor by political pundits and the press since he surpassed his "expected" vote. In such situations, expectations become the benchmark against which reality is tested. See Richard M. Scammon and Ben J. Wattenberg, *The Real Majority: An Extraordinary Examination of the American Electorate* (New York: Coward, McCann and Geohagen, 1970), pp. 27, 85-93.
5. Strictly speaking, it is not necessary to postulate winning as the most-preferred outcome. Rather, "rationality" may be defined in terms of the choice of the *most valued* outcome, where the value associated with winning coalitions and the payoffs to their members are stipulated by assumptions 5 and 6.
6. William H. Riker and Peter C. Ordeshook, *An Introduction to Positive Political Theory* (Englewood Cliffs, N.J.: Prentice-Hall, 1973), pp. 179-180; for further details, see Riker, *Theory of Political Coalitions,* pp. 40-46, 247-278.
7. Riker, *Theory of Political Coalitions,* p. 47.
8. Riker and Ordeshook, *Introduction to Positive Political Theory,* pp. 194-196. Whether or not Johnson was conscious beforehand of the effects his actions would have is irrelevant to the test of the size principle, which asserts that he could not have prevented the breakup of his grand coalition no matter how hard he tried. That is, the game-theoretic logic of the model says that some of his excess supporters would invariably have been disaffected and attracted to

the opposition coalition, whomever he tried to please with his policies. That Johnson was aware of the dilemma he faced is suggested by the following remarks he made to Doris Kearns:

> "I knew from the start," Johnson told me in 1970, describing the early weeks of 1965, "that I was bound to be crucified either way I moved. If I left the woman I really loved—the Great Society—in order to get involved with that bitch of a war on the other side of the world, then I would lose everything at home. All my programs. All my hopes to feed the hungry and shelter the homeless. All my dreams to provide education and medical care to the browns and the blacks and the lame and the poor. But if I left that war and let the Communists take over South Vietnam, then I would be seen as a coward and my nation would be seen as an appeaser and we would both find it impossible to accomplish anything for anybody anywhere on the entire globe." (Doris Kearns, *Lyndon Johnson and the American Dream* [New York: New American Library, 1977], p. 263)

9. Much of this literature is critically reviewed in Steven J. Brams, *Game Theory and Politics* (New York: Free Press, 1975), pp. 226-232. See also Eric C. Browne and John Dreijmanis, eds., *Government Coalitions in Western Democracies* (New York: Longman, 1982); James P. Kahan and Amnon Rapoport, *Theories of Coalition Formation* (Hillsdale, N.J.: Lawrence Erlbaum Associates, 1984); and Philip D. Straffin, Jr., and Bernard Grofman, "Parliamentary Coalitions: A Tour of Models," *Mathematics Magazine* 57 (November 1984): 259-274.
10. Riker, *Theory of Political Coalitions,* pp. 107-108.
11. Ibid., pp. 149-158.
12. If a deadlock had developed, this appears indeed to have been their intention. See John Spencer Bassett, *The Life of Andrew Jackson,* 2d ed. (New York: Macmillan, 1925), p. 368.
13. I have assumed that Clay made the first commitment not only because as low man he was eliminated from the contest in the House but also because Crawford's supporters felt bound to vote for their candidate on the first ballot. Bassett, *Life of Andrew Jackson,* p. 363.
14. "Minimal winning" here means exactly a bare majority, whereas in Section 5.4 a minimal winning coalition (MWC) is a coalition in which the defection of at least one member—when members might have different weights—would cause it to be losing.
15. Bassett, *Life of Andrew Jackson,* pp. 350-351.
16. Ibid., p. 352.
17. Marquis James, *Andrew Jackson: Portrait of a President* (New York: Grosset and Dunlap, 1937), p. 120
18. Riker views skeptically an interesting (and possibly apocryphal) story about the intense pressure exerted on this representative and how he made his decision. Riker, *Theory of Political Coalitions,* pp. 155-157.
19. Although the calculations are rather complicated, some attempt at generalization has been made through the utilization of lattice structures for arraying the step-by-step buildup of protocoalitions in 10-member voting bodies and

computer-simulation techniques for larger bodies. See Steven J. Brams and William H. Riker, "Models of Coalition Formation in Voting Bodies," in *Mathematical Applications in Political Science,* vol. 6, ed. James F. Herndon and Joseph L. Bernd (Charlottesville: University Press of Virginia, 1972), pp. 79-124; Steven J. Brams, "A Cost/Benefit Analysis of Coalition Formation in Voting Bodies," in *Probability Models of Collective Decision Making,* ed. Richard G. Niemi and Herbert F. Weisberg (Columbus, Ohio: Charles E. Merrill, 1972), pp. 101-124; Steven J. Brams and John G. Heilman, "When to Join a Coalition, and with How Many Others, Depends on What You Expect the Outcome to Be," *Public Choice* 17 (Spring 1974): 11-25; and P. D. Straffin, Jr., M. D. Davis, and S. J. Brams, "Power and Satisfaction in an Ideologically Divided Voting Body," in *Power, Voting, and Voting Power,* ed. Manfred J. Holler (Würzburg, West Germany: Physica-Verlag, 1982), pp. 239-255.

20. Steven J. Brams and José E. Garriga-Picó, "Bandwagons in Coalition Formation: The 2/3's Rule," *American Behavioral Scientist* 18 (March/April 1975): 472-496. This article is reprinted in Barbara Hinckley, ed., *Coalitions and Time: Cross-Disciplinary Studies* (Beverly Hills, Calif.: Sage Publications, 1976), pp. 34-58.
21. Philip D. Straffin, Jr., "The Bandwagon Curve," *American Journal of Political Science* 21 (November 1977): 695-709.
22. See L. S. Shapley and Martin Shubik, "A Method of Evaluating the Distribution of Power in a Committee System," *American Political Science Review* 48 (September 1954): 787-792, for a description of the Shapley-Shubik index.
23. Brams and Riker, "Models of Coalition Formation in Voting Bodies."
24. Straffin, "Bandwagon Curve," p. 700.
25. Ibid., p. 702.
26. Ibid., p. 703.
27. Thomas Romer and Howard Rosenthal, "Voting Models and Empirical Evidence," *American Scientist* 72 (September/October 1984): 472.

9 Strategy and Ethics

The rational-choice models in political science surveyed here affirm that political actors, for plausible assumptions about their goals, act rationally. They choose better over worse alternatives in light of their preferences and the preferences of other actors, or players, in games.

Politics is distinguished from other human activities by the fact that the preferences of players conflict, and the choices they make have significant consequences for actors other than themselves. It is manifest in the Bible, as when Esther and Mordecai thwarted Haman's planned genocide of the Jews and averted a possible calamity that would have gone well beyond them. Similarly, when candidates seek to outmaneuver their competitors by offering voters greater expected rewards; when legislators try to manipulate the agenda or vote strategically to try to secure better outcomes; and when nation-states and their leaders endeavor to avoid disastrous wars through threats, the exercise of power, or the formation of alliances, their actions have larger ramifications.

The strategies players settle upon for pursuing their goals are generally carefully thought out—if not always successful in achieving these goals. Yet, despite the best efforts of players in political games, it may not be possible for them to realize favorable outcomes. For example, President Nixon to his chagrin was forced to resign, and the costly arms race between the superpowers has persisted, though the players in these games would surely have preferred different outcomes. But the nature of traps often condemns players, even wily ones, to seemingly irrational outcomes.

An outcome that hurts all players, however, is not unequivocal evidence that the players are irrational, particularly if this outcome is the product of rational strategy choices, as in prisoners' dilemma. Quite the contrary: the choice by players of their dominant, or unconditionally best, strategies in games like prisoners' dilemma is eminently rational, though sometimes an alteration of the rules of such intractable games may permit the implementation of stable and more satisfactory outcomes.

If less-than-satisfactory outcomes for the players of certain games occur, and no player can improve his or her payoff by switching strategies, rational politics would appear to be problematic from a normative viewpoint. But even in situations that are not traps, the morality of strategic choices is not always evident. Take, for example, the story of Esther. Her actions seem largely to have been based on calculations of private gain—including the saving of her own life—rather than of preventing genocide. Nevertheless, her apparently selfish behavior had massively beneficial consequences for all Jews, so it is hard to condemn her behavior even if her motives are suspect. After all, most of us act to satisfy rather personal goals; fortunately, only very rarely are the consequences catastrophic for others.

The choices of political figures are more consequential because by definition they affect more people. The morality of politicians' goals may be harder to assess, for, typically, their rational choices in pursuit of these goals hurt some while helping others.

President Nixon's resignation, coming at the time it did, probably was a net benefit to the country because it was spared a bitter and divisive trial by the Senate. The choices made both by Nixon and by Burger and Blackmun in the White House tapes case, if rational, were nevertheless costly for them. Because the two Supreme Court justices voted against their apparent convictions, they too paid a price; moreover, their behavior would also seem unethical—except as judged against the fact that the game they played left them with no recourse.

Ethics, in other words, seems best viewed in light of the situation (or game, in this case) in which these choices are enmeshed (for which the appelation "situational ethics" is sometimes used). The decision of some of the justices to forsake their convictions to make the Supreme Court's voice effective—by making defiance by Nixon extremely difficult—parallels Esther's decision to save her life by saving her people. In each case, these decisions were wrought of what the strategic situation permitted the players to do—at the moment—to accomplish their goals.

The actors' goals in these situations seemed worthy ones, and their rational actions were certainly supportive. One could offer a similar defense of Solidarity's goals in the Polish crisis, or of the goals of the United States in the Cuban missile crisis. Emphatically, however, there is nothing *inherently* amoral about the goals of the other sides—the Communist party in Poland or the Soviet Union in the Cuban missile crisis—as mirrored by their preferences in the games analyzed. Nor is Haman's goal of preserving his own favored position an unreasonable one. Yet the means he attempted to achieve it almost succeeded in bringing about a terrible tragedy. Although Nixon did not threaten to wreak havoc on a whole society—only on his "enemies"—like Haman he used what are generally regarded as ignoble means in pursuit of his foes.

Whether successful or not, the players in these various games acted rationally, given their preferences and the powers each had. Quite apart from their rationality, however, the moral approbation or disapprobation one attaches to the resulting outcomes is a subjective judgment, though it may be rooted in a system of ethics. Consider the "corrupt bargain" of 1825. Can it be condoned because it furthered John Quincy Adams's goal of winning the presidency? More generally, can payoffs or even bribes be justified if one sees an overriding value in ensuring, say, the election of one person over a dreaded opponent?

Although moral judgments about game outcomes may reflect which player one favors, this need not always be the case. For example, one might applaud the selection of the compromise outcomes in prisoners' dilemma or chicken, even though they are not the best for either player.

If, as an ethical tenet, compromise outcomes in difficult games like prisoners' dilemma and chicken deserve moral approval, the rules of games, insofar as this is possible, should be altered to make the rational choice of such outcomes strategically more defensible. Similarly, if there are good ethical reasons to prefer certain social choices, such as Condorcet candidates (if they exist), over others, election reforms should be encouraged that make the selection of these social choices more probable and that also avoid, insofar as possible, some of the voting paradoxes. Likewise with apportionment: certain systems lead to fewer anomalies than others and hence deserve more serious consideration.

To be sure, there will never be a consensus on what are the most desirable attributes of either voting or apportionment systems. This divergence is natural, because some systems benefit certain parties, other systems other parties. Nevertheless, models of social choice unquestionably help one distinguish what properties different systems satisfy and what the trade-offs among them are. Thereby one can make a more intelligent choice of what is best for one's purposes and hence should be promoted, which is itself a rational choice—if not of a strategy in a game, then of a system consonant with principles one considers palatable, even laudable.

Although one can argue for some principles over others if they are in conflict, ultimately one faces a value decision in making choices. In my opinion, it is better to have an understanding of what values are at stake, which rational-choice models can clarify, than to engage in a fruitless debate over the oft-touted virtues of democracy.

For example, should order take precedence over justice, or justice over order? The vagaries of traditional democratic theory do not enable one to determine what trade-offs in values one faces when the debate is couched in terms of abstract and protean concepts like order and justice, unrelated to a particular context provided by the assumptions of a model.

I have devoted considerable space to the analysis of candidate and voter strategies in a democracy and have given a number of examples of each. Also, I have looked at coalitions that are likely to form and remain stable in American and international politics, and the effects of power in different games. If the copious results that come from this kind of analysis cannot be easily summarized, I believe they have a bearing, nonetheless, on making better ethical judgments and avoiding wrenching predicaments.

As I attempted to show in several games, the ethics of rational choice is not dependent only on the presumed morality of goals. When mediated by particular situational constraints in a rational-choice model, goals may have a rather different meaning than when considered in the abstract. In particular, when the means to satisfy goals are brought into the picture in any situation, they lend an additional dimension to normative analysis: they force the ethicist to consider goals in light of the rational means used to achieve them.

This linking of means and ends in rational-choice analysis does not answer the question of what is right or wrong. Instead, it makes more perspicuous what goals entail in terms of rational choices in a particular situation. Thereby this kind of analysis can supplement ethical evaluation by showing up logical relationships among means and ends that facilitate making more rounded normative judgments.

Put another way, the question of right and wrong is easier to address in a particular situation than in terms of some ethical abstraction. For then one can evaluate, with the aid of rational-choice models, the means and ends simultaneously.

The conjunction of means and ends requires more refined philosophical analysis than I can give it here. Nevertheless, it seems clear from the examples considered that ethics and strategy are inseparable, each enlarging the other. For just as ethical evaluation benefits from the strategic analysis of rational politics, so, too, rational theorists benefit from tackling ethical issues that their analysis helps to raise.

This is so, I believe, because strategic considerations can illuminate moral issues, such as whether there is an inherent rightness one can attach to social choices.[1] Similarly, ethics can augment strategic issues by helping one better understand, foresee, and thus avoid moral dilemmas, such as those implicit in several of the games studied. A more strategically defensible ethics is not, I think, an insignificant byproduct of the study of rational politics.

NOTE

1. This question is analyzed in William H. Riker, *Liberalism against Populism: A Confrontation between the Theory of Democracy and the Theory of Social Science* (San Francisco: Freeman, 1982), and Nicholas R. Miller, "Pluralism and Social Choice," *American Political Science Review* 77 (September 1983): 734-747.

Glossary

This glossary contains definitions of game-theoretic and other concepts used in this book. Specific games and concepts that have been developed in detail in the text or that would require extended discussion (for example, the Banzhaf power index) are not included. As in the text, an attempt has been made to define these terms in relatively nontechnical language.

Adjusted district voting. System of proportional representation in which parties that are underrepresented from single-member district elections may be given additional seats, based on their national proportions, resulting in a variable-size legislature.

Amendment procedure. Voting procedure under which an amendment is voted on before a bill; the final vote is on passage of the resulting outcome.

Approval voting. Voting system in which voters can vote for as many candidates as they like, or approve of, in an election with more than two candidates.

Backward induction. Reasoning process in which players, working backward from the bottom to the top of a game tree, anticipate each other's rational choices.

Bandwagon curve. Defines the region in which it is rational for an uncommitted actor to join the larger of two protocoalitions as a function of the respective sizes of the protocoalitions.

Borda count. System of preferential voting in which each voter's first-choice candidate gets the most points and lower choices fewer points; the candidate with the most points wins.

Choice rule. In a two-person game, a conditional strategy based on one player's prediction of the strategy choice of the other.

Coalition. A subset of players that is formed to achieve some end mutually beneficial to its members.

Compellent threat. In repeated play of a sequential two-person game, one player's (the threatener's) threat to stay at a particular strategy to induce the threatened player to choose its (as well as the threatener's) best outcome associated with that strategy.

Conditional cooperation. In a two-person game, a choice rule of tit-for-tat that says that a player will cooperate if he or she predicts that the other player will cooperate; otherwise he or she will not.

Condorcet candidate. A candidate who can defeat all others in a series of pairwise contests.

Connectivity. Property of a voter's preferences if, for every pair of alternatives, he or she either strictly prefers one to the other or is indifferent.

Constant-sum (zero-sum) game. A game in which the payoffs to the players at every outcome sum to some constant (or zero); all constant-sum games can be transformed into zero-sum games by subtracting the appropriate constant from the payoffs to the players.

Critical defection. In a weighted-majority game, a defection that transforms a minimal winning coalition (with respect to at least one player) into a losing coalition.

Cyclical majorities. Formed when majorities of voters prefer candidate a to b, b to c, and c to a, indicating the lack of a social choice or consensus; such majorities exist when there is a paradox of voting.

Deception strategy. In a game of incomplete information, a player's false announcement of his or her preferences to induce the other player to choose a strategy favorable to the deceiver.

Decision rule. In a voting body, specifies the number of votes required to take collective action that is binding on all the members.

Deterrent threat. In repeated play of a sequential two-person game, a threat to move to another strategy to induce the threatened player to choose an outcome, associated with the threatener's initial strategy, that is better for both players than the outcomes threatened.

Dictator. In a weighted-majority game, a player whose votes are sufficient to determine the outcome of the voting game.

Dominant strategy. One that leads to outcomes at least as good as any other strategy for all possible choices of other players and to a better outcomes for at least one set of choices.

Dominated strategy. One that leads to outcomes no better than those given by any other strategy for all possible choices of other players and to a worse outcome for at least one set of choices.

Dummy. In a weighted-majority game, a player whose votes never have any effect on the outcome of the game.

Effective power. In a two-person game, power that enables a player to induce a different (and usually better) outcome for himself or herself when he or she possesses this power than when the other player possesses it.

Equilibrium. See Nash equilibrium; Nonmyopic equilibrium.

Expected payoff. The sum of the payoffs a player receives from each outcome, multiplied by the outcome's probability of occurrence, for all possible outcomes that may arise.

Final outcome. In a sequential game, the outcome induced by (possibly) rational moves and countermoves from the initial outcome, according to the theory of moves.

Game. May take different forms (as indicated in subsequent *game* entries); more generally, the totality of rules of play that describe a strategic situation.

Game of complete information. One in which the players know each others' preferences or payoffs for every outcome and the rules of play.

Game in extensive form. One represented by a game tree in which players are assumed to make sequential choices.

Game in normal form. One represented by a payoff matrix in which players are assumed to make independent strategy choices.

Game of partial conflict. A variable-sum game in which the players' preferences are not diametrically opposed.

Game of perfect information. One in which each player knows with certainty the strategy choice or move of every other player at each point in the sequence of play.

Game theory. Mathematical theory of strategy to explicate optimal choices in interdependent decision situations, wherein the outcome depends on the choices of two or more actors, or players.

Game of total conflict. A constant-sum game in which what one player gains the other players lose.

Game tree. Symbolic tree based on the rules of play of the game, in which the vertices, or nodes, of the tree represent choice points and the branches represent alternative courses of action that can be selected.

Grand coalition. One that contains all the players in a game.

Hare system. System of preferential voting, also known as "single transferable vote," under which if no candidate receives a majority of first-place votes, the candidate with the fewest first-place votes is dropped and his or her second-place votes are given to the remaining candidates. This elimination process continues, with lower-place votes of the voters whose preferred candidates are eliminated being transferred to the candidates that survive, until one or more candidates receives a prespecified quota (simple majority if only one candidate is to be chosen).

Information effect. The tendency of incomplete or imperfect information to induce coalitions to form that are larger than minimal winning size as a cushion against uncertainty.

Initial oucome. In a sequential game, the outcome rational players choose when they make their initial strategy choices according to the theory of moves.

Median. The point at which a line drawn through a voter distribution curve divides the area under the curve exactly in half.

Minimal winning coalition. Coalition that would be rendered losing by the defection of at least one of its members.

Monotonicity. Property of a preferential voting system indicating that a candidate cannot be hurt by being raised in the preference rankings of some voters while remaining the same in the rankings of all others.

Move. In a game in extensive form, a choice point, or node, of the game tree where one from a given set of alternative courses of action is chosen by a player.

Moving power. In a two-person sequential game, the ability to continue moving when the other player must eventually stop.

Nash equilibrium. In a normal-form game, an outcome from which no player would have an incentive to depart unilaterally because he or she would do (immediately) worse, or at least not better, by departing.

Nonmyopic calculation. In a two-person sequential game, choices made by rational players in full anticipation of how each will respond to the other, both in selecting their strategies initially and making subsequent moves.

Nonmyopic equilibrium. In a two-person sequential game, an outcome from which neither player, anticipating all possible rational moves and countermoves from the initial outcome, would have an incentive to depart, because he or she would do (eventually) worse, or at least not better, by doing so.

Ordinal game. One in which the players can rank, but not necessarily assign payoffs or utilities to, the outcomes.

Paradox of the chair's position. Occurs when being chair (with a tie-breaking vote) hurts rather than helps the chair if voting is sophisticated.

Paradox of new members. Occurs when the addition of one or more players to a weighted-voting body increases the voting power of at least one of the original players.

Paradox of voting. Occurs when no alternative can defeat all others in a series of pairwise contests if voting is sincere.

Pareto-inferior outcome. Outcome in game wherein there exists another outcome that is better for some players and not worse for all the others.

Pareto-superior outcome. Outcome in game that is better for some players than, or as good for all players as, any other outcome.

Payoff. Utility, or numerical value, that a player receives at an outcome in a game.

Payoff matrix. Rectangular array, or matrix, whose entries indicate the payoffs to players resulting from their strategy choices at every outcome of the game.

Pivot. Player who is decisive to a protocoalition's becoming winning in a voting game as players are successively added to form the grand coalition.

Player. See Rational player.

Plurality voting. Voting system in which voters can vote for only one candidate.

Population monotonicity. Property of an apportionment system in which a state's gaining in population can never result in its losing a seat when the total population remains fixed.

Preferences. A player's ranking of outcomes from best to worst.

Protocoalition. A coalition that is not sufficiently large to be winning.

Public bad. See Public good.

Public good. A good that, when supplied to some members of the public, cannot be withheld from other members; it may benefit (public good) or not benefit (public bad) the members.

Quota. In an apportionment system, the exact number (whole number plus fractional remainder) of seats a state is entitled to; an apportionment system satisfies quota if states always receive this number either rounded up or rounded down.

Rational player. One who seeks to attain better outcomes, according to his or her preferences, in light of the presumed rational choices of other players in a game.

Revealed deception. In a game of incomplete information, tactic of a player's falsely announcing one strategy to be dominant but subsequently choosing another strategy, thereby revealing deception.

Rules of play. Describe the preferences and choices available to the players, their sequencing, and any special prerogatives (for example, power) one player may have in the play of the game.

Sequential game. One in which players can move and countermove, in accordance with the theory of moves, after their initial strategy choices.

Share of spoils. An expected value equal to the probabilistic contribution an actor makes to a protocoalition times the probability that, with this contribution, the protocoalition will go on to win.

Side payments. Payments involving some common medium of exchange, such as money, that can be transferred among the players before or after play in an n-person game.

Sincere voting. Voting directly in accordance with one's preferences.

Single-peakedness. In reference to the shape of voter preferences, occurring when there exists a single dimension underlying the preferences of voters (for example, a liberalism-conservatism scale) along which alternatives can be ordered; it precludes the existence of a paradox of voting.

Size principle. Principle that only minimal winning coalitions will form in n-person zero-sum games in which players are permitted to make side payments, are rational, and have complete and perfect information.

Sophisticated voting. Voting that forecloses the possibility that a voter's worst outcomes will be chosen—insofar as this is possible—through the successive elimination of dominated strategies, given that other voters act likewise.

Staying power. In a two-person sequential game, the ability of a player to hold off making a strategy choice until the other player has made his or hers.

Strategic voting. Voting that is not sincere and that is intended to bring about preferred outcomes.

Strategy. In a game in normal form, a complete plan that specifies all possible courses of action of a player for whatever contingencies may arise.

Strategyproofness constraint. Constraint under adjusted district voting that prevents a party from gaining more seats when it loses rather than wins a district election.

Symmetrical game. Two-person game in which the ranks of the outcomes by the players along the main diagonal are the same, whereas the ranks of the off-diagonal outcomes are mirror images.

Tacit deception. In a game of incomplete information, a player's falsely announcing one strategy to be dominant, and subsequently choosing this strategy, so as not to reveal his or her deception.

Theory of moves. Describes optimal strategic calculations in normal-form games in which the players can move and countermove from an initial outcome in sequential play.

Threat power. In a two-person sequential game that is repeated, the ability of a player to threaten a mutually disadvantageous outcome in the single play of a game to deter untoward actions in the future play of this or other games.

Tit-for-tat. See Conditional cooperation.

Transitivity. If a voter prefers a to b and b to c, he or she will prefer a to c. Transitivity is also applicable to social-choice orderings of alternatives; when these orderings are not transitive, majorities are said to be cyclical.

Traps. Games in which players' apparently rational strategies lead them to a collectively worse (that is, Pareto-inferior) outcome than had they chosen other strategies.

Undominated strategy. One that is neither unconditionally best, or dominant, nor unconditionally worst, or dominated.

Unimodal voter distribution. Voter distribution curve with one peak, or mode.

Value-restrictedness. Property of a set of preference scales such that all actors agree that there is some alternative which is never best, medium, or worst for every set of three alternatives, or triple.

Variable-sum game. One in which the sum of the payoffs to the players at different outcomes varies, so the players may gain or lose simultaneously at different outcomes.

Weighted-majority game. Voting game defined by players of specified weights and a decision rule (for example, simple majority) that distinguishes between coalitions of winning and losing size.

Zero-sum game. See Constant-sum (zero-sum) game.

Bibliography

Abel, Elie. *The Missile Crisis.* Philadelphia: Lippincott, 1966.

Abrams, Robert. *Foundations of Political Analysis: An Introduction to the Theory of Collective Choice.* New York: Columbia University Press, 1980.

Allen, Thomas B., and Norman Polmar. "The Silent Chase: Tracking Soviet Submarines." *New York Times Magazine,* January 1, 1984, pp. 13-17, 26-27.

Allison, Graham T. *Essence of Decision: Explaining the Cuban Missile Crisis.* Boston: Little, Brown, 1971.

Arrington, Theodore S., and Saul Brenner, "Another Look at Approval Voting." *Polity* 17 (Fall 1984): 188-134.

Arrow, Kenneth J. *Social Choice and Individual Values.* 2d ed. New Haven, Conn.: Yale University Press, 1963.

Ascherson, Neal. *The Polish August: The Self-Limiting Revolution.* New York: Viking, 1982.

Aspin, Les. "The Verification of the SALT II Agreement." *Scientific American,* February 1979, pp. 38-45.

Axelrod, Robert. *The Evolution of Cooperation.* New York: Basic, 1984.

Balinski, Michel L., and H. Peyton Young. *Fair Representation: Meeting the Ideal of One Man, One Vote.* New Haven, Conn.: Yale University Press, 1982.

Banzhaf, John F., III. "Weighted Voting Doesn't Work: A Mathematical Analysis." *Rutgers Law Review* 19 (Winter 1965): 317-343.

Barry, Brian, and Russell Hardin, eds. *Rational Man and Irrational Society: An Introduction and Sourcebook.* Beverly Hills, Calif.: Sage, 1982.

Bassett, John Spencer. *The Life of Andrew Jackson.* New York: Macmillan, 1925.

Bell, Roderick, David V. Edwards, and R. Harrison Wagner, eds. *Political Power: A Reader in Theory and Research.* New York: Free Press, 1969.

Bialer, Severyn. "Poland and the Soviet Imperium." *Foreign Affairs* 59 (1981): 522-539.

Biddle, W. F. *Weapons, Technology, and Arms Control.* New York: Praeger, 1972.

Black, Duncan. *Theory of Committees and Elections.* Cambridge: Cambridge University Press, 1958.

Blydenburgh, John C. "The Closed Rule and the Paradox of Voting." *Journal of Politics* 33 (February 1971): 57-71.

Bogdanor, Vernon. *The People and the Party System: The Referendum and Electoral Reform in British Politics.* Cambridge: Cambridge University Press, 1981.

——. *What Is Proportional Representation? A Guide to the Issues.* Oxford: Martin Robertson, 1984.

Bone, Hugh A., and Austin Ranney. *Politics and Voters.* New York: McGraw-Hill, 1976.

Bordley, Robert F. "A Pragmatic Method for Evaluating Election Schemes through Simulation." *American Political Science Review* 72 (September 1983): 831-847.

Boulding, Kenneth E., ed. *Peace and the War Industry.* New Brunswick, N.J.: Transaction, 1973.

Bowen, Bruce D. "Toward an Estimate of the Frequency of the Paradox of Voting in U.S. Senate Roll Call Votes." In *Probability Models of Collective Decision Making,* edited by Richard G. Niemi and Herbert F. Weisberg. Columbus, Ohio: Charles E. Merrill, 1972, pp. 181-203.

Brams, Steven J. "The AMS Nomination Procedure Is Vulnerable to 'Truncation of Preferences.'" *Notices of the American Mathematical Society* 29 (February 1982): 136-138.

——. *Biblical Games: A Strategic Analysis of Stories in the Old Testament.* Cambridge, Mass: MIT Press, 1980.

——. "A Cost/Benefit Analysis of Coalition Formation in Voting Bodies." In *Probability Models of Collective Decision Making,* edited by Richard G. Niemi and Herbert F. Weisberg. Columbus, Ohio: Charles E. Merrill, 1972, pp. 101-124.

——. "Deception in 2 x 2 Games." *Journal of Peace Science* 2 (Spring 1977): 171-203.

——. *Game Theory and Politics.* New York: Free Press, 1975.

——. "Newcomb's Problem and Prisoners' Dilemma." *Journal of Conflict Resolution* 19 (December 1975): 596-612.

——. "Omniscience and Omnipotence: How They May Help—or Hurt—in a Game." *Inquiry* 25 (June 1982): 217-231.

——. *Paradoxes in Politics: An Introduction to the Nonobvious in Political Science.* New York: Free Press, 1976.

——. *The Presidential Election Game.* New Haven, Conn.: Yale University Press, 1978.

——. *Spatial Models of Election Competition.* Monographs in Undergraduate Mathematics and Its Application. Newton, Mass.: Education Development Center, 1979; reprint, Lexington, Mass.: COMAP, 1983.

——. "Strategic Information and Voting Behavior." *Society* 19 (September/October 1982): 4-11.

——. *Superior Beings: If They Exist, How Would We Know? Game-Theoretic Implications of Omniscience, Omnipotence, Immortality, and Incomprehensibility.* New York: Springer-Verlag, 1983.

——. *Superpower Games: Applying Game Theory to Superpower Conflict.* New Haven, Conn.: Yale University Press, 1985.

Brams, Steven J., and Paul J. Affuso. "New Paradoxes of Voting Power on the EC Council of Ministers." *Electoral Studies* 4 (August 1985): 187-191.

——. "Power and Size: A New Paradox." *Theory and Decision* 7 (February/May 1976): 29-56.

Brams, Steven J., and Morton D. Davis. "The 3/2's Rule in Presidential

Campaigning." *American Political Science Review* 68 (March 1974): 113-134.

——. "The Verification Problem in Arms Control: A Game-Theoretic Analysis." In *Interaction and Communicaton in Global Politics,* edited by Claudio Cioffi-Revilla, Richard L. Merritt, and Dina A. Zinnes. Beverly Hills, Calif.: Sage 1985.

Brams, Steven J., Morton D. Davis, and Philip D. Straffin, Jr. "Communications" (comment on R. H. Wagner). *American Political Science Review* 78 (June 1984): 495.

——. "The Geometry of the Arms Race." *International Studies Quarterly* 23 (December 1979): 567-588.

——. "A Reply to 'Detection and Disarmament.'" *International Studies Quarterly* 23 (December 1979): 599-600.

Brams, Steven J., Dan S. Felsenthal, and Zeev Maoz. "Chairman Paradoxes under Approval Voting." Mimeographed, 1985.

——. "New Chairman Paradoxes." Mimeographed, 1985.

Brams, Steven J., and Peter C. Fishburn. *Approval Voting.* Boston: Birkhäuser, 1983.

——. "A Careful Look at 'Another Look at Approval Voting.'" *Polity* 17 (September 1984): 135-143.

——. "Communications" (comment on R. G. Niemi). *American Political Science Review* 79 (September 1985).

——. "A Note on Variable-Size Legislatures to Achieve PR." In *Choosing an Electoral System: Issues and Alternatives,* edited by Bernard Grofman and Arend Lijphart. New York: Praeger, 1984, pp. 175-177.

——. "Proportional Representation in Variable-Size Legislatures." *Social Choice and Welfare* 1 (1984): 211-229.

——. "Some Logical Defects of the Single Transferable Vote." In *Choosing an Electoral System: Issues and Alternatives,* edited by Bernard Grofman and Arend Lijphart. New York: Praeger, 1984, pp. 147-151.

Brams, Steven J., and José E. Garriga-Picó. "Bandwagons in Coalition Formation: The 2/3's Rule." *American Behavioral Scientist* 18 (March/April 1975): 472-496. Reprinted in *Coalitions and Time: Cross-Disciplinary Studies,* edited by Barbara J. Hinckley. Beverly Hills, Calif.: Sage, 1976, pp. 34-58.

Brams, Steven J., and John G. Heilman. "When to Join a Coalition, and with How Many Others, Depends on What You Expect the Outcome to Be." *Public Choice* 17 (Spring 1974): 11-25.

Brams, Steven J., and Marek P. Hessel. "Absorbing Outcomes in 2 x 2 Games." *Behavioral Science* 27 (October 1982): 393-401.

——. "Staying Power in 2 x 2 Games." *Theory and Decison* 15 (September 1983): 279-302.

——. "Threat Power in Sequential Games." *International Studies Quarterly* 28 (March 1984): 23-44.

Brams, Steven J., and Douglas Muzzio. "Game Theory and the White House Tapes Case." *Trial* 13 (May 1977): 48-53.

——. "Unanimity in the Supreme Court: A Game-Theoretic Explanation of the Decision in the White House Tapes Case." *Public Choice* 32 (Winter 1977): 67-83.

Brams, Steven J., and William H. Riker. "Models of Coalition Formation in Voting Bodies." In *Mathematical Applications in Political Science, VI,* edited by James F. Herndon and Joseph L. Bernd. Charlottesville: University Press of Virginia, 1972, pp. 79-124.

Brams, Steven J., and Philip D. Straffin, Jr. "The Apportionment Problem." *Science,* July 30, 1982, pp. 437-438.

——. "The Entry Problem in a Political Race." In *Political Equilibrium,* edited by Peter C. Ordeshook and Kenneth A. Shepsle. Boston: Kluwer Nijhoff, 1982, pp. 181-195.

Brams, Steven J., and Donald Wittman. "Nonmyopic Equilibria in 2 x 2 Games." *Conflict Management and Peace Science* 6 (Fall 1981): 39-62.

Brams, Steven J., and Frank C. Zagare. "Deception in Simple Voting Games." *Social Science Research* 6 (September 1977): 257-272.

——. "Double Deception: Two against One in Three-Person Games." *Theory and Decision* 13 (March 1981): 81-90.

Browne, Eric C., and John Dreijmanis, eds. *Government Coalitions in Western Democracies.* New York: Longman, 1982.

Bueno de Mesquita, Bruce. *The War Trap.* New Haven, Conn.: Yale University Press, 1981.

Chamberlin, John R., and Paul R. Courant. "Representative Deliberation and Representative Decisions: Proportional Representation and the Borda Rule." *American Political Science Review* 77 (December 1983): 718-733.

Chayes, Abram. *The Cuban Missile Crisis: International Crises and the Role of Law.* New York: Oxford University Press, 1974.

Clarke, John H. "Judicial Power to Declare Legislation Unconstitutional." *American Bar Association Journal* 9 (November 1923): 689-692.

Coleman, James S. "Communications." *American Political Science Review* 67 (June 1973): 567-569.

Cox, Gary W. "Electoral Equilibria in Multicandidate Elections: Plurality versus Approval Voting." *American Journal of Political Science* 29 (February 1985): 112-118.

Cross, John G., and Melvin J. Guyer. *Social Traps.* Ann Arbor: University of Michigan Press, 1980.

Dacey, Raymond. "Detection and Disarmament: A Comment on 'The Geometry of the Arms Race.'" *International Studies Quarterly* 23 (December 1979): 589-598.

——. "Detection, Inference and the Arms Race." In *Reason and Decision,* edited by Michael Bradie and Kenneth Sayre. Bowling Green Studies in Applied Philosophy, vol. 3 (1981). Bowling Green, Ohio: Applied Philosophy Program, Bowling Green State University, 1982, pp. 87-100.

Daniel, Donald C., and Katherine L. Herbig, eds. *Strategic Military Deception.* New York: Pergamon, 1982.

Detzer, David. *The Brink: Story of the Cuban Missile Crisis.* New York: Crowell, 1979.

Dinerstein, Herbert. *The Making of the Cuban Missile Crisis, October 1962.* Baltimore: Johns Hopkins University Press, 1976.

Divine, Robert A., ed. *The Cuban Missile Crisis.* Chicago: Quadrangle, 1971.

Doron, Gideon, and Richard Kronick. "Single Transferable Vote: An Example of a Perverse Social Choice Function." *American Journal of Political Science* 21 (May 1977): 303-311.

Downs, Anthony. *An Economic Theory of Democracy.* New York: Harper and Row, 1957.

Dummett, Michael. *Voting Procedures.* Oxford: Oxford University Press, 1984.

Enelow, James M., and Melvin J. Hinich. *The Spatial Theory of Election Competition: An Introduction.* Cambridge: Cambridge University Press, 1984.

"Enlargement of the Community: Transition Period and Institutional Implications." *Bulletin of the European Communities* 11 (Supplement, February 1978).

Evans, Roland, and Robert Novak. "Mr. Nixon's Supreme Court Strategy." *Washington Post,* June 12, 1974, p. A29.

Farquharson, Robin. *Theory of Voting.* New Haven, Conn.: Yale University Press, 1969.

Feldman, Allan M. *Welfare Economics and Social Choice Theory.* Boston: Martinus Nijhoff, 1980.

Fishburn, Peter C. "Monotonicity Paradoxes in the Theory of Elections." *Discrete Applied Mathematics* 4 (April 1982): 119-134.

——— . *The Theory of Social Choice.* Princeton, N.J.: Princeton University Press, 1973.

Fishburn, Peter C., and Steven J. Brams. "Manipulability of Voting by Sincere Truncation of Preferences." *Public Choice* 44, no. 3 (1984): 397-410.

——— . "Paradoxes of Preferential Voting." *Mathematics Magazine* 56 (September 1983): 207-214.

The Five Megilloth and Jonah: A New Translation. Philadelphia: Jewish Publication Society of America, 1969; 2d rev. ed., 1974.

Fraser, Niall M., and Keith W. Hipel. "Dynamic Modeling of the Cuban Missile Crisis." *Conflict Management and Peace Science* 6 (Spring 1982-1983): 1-18.

Frohlich, Norman, and Joe A. Oppenheimer. *Modern Political Economy.* Englewood Cliffs, N.J.: Prentice-Hall, 1978.

Gamson, William A., and André Modigliani. *Untangling the Cold War: A Strategy for Testing Rival Theories.* Boston: Little, Brown, 1971.

Garthoff, Raymond L. "The Role of Nuclear Weapons: Soviet Perceptions." In *Nuclear Negotiations: Reassessing Arms Control Goals in U.S.-Soviet Relations,* edited by Alan F. Neidle. Austin, Texas: Lyndon B. Johnson School of Public Affairs, 1982, pp. 10-13.

Gibbard, Alan. "Manipulation of Voting Schemes: A General Result." *Econometrica* 41 (May 1973): 587-601.

Gillespie, John V., Dina A. Zinnes, G. S. Tahim, Philip A. Schrodt, and R. Michael Robison. "An Optimal Control Model of Arms Races." *American Political Science Review* 71 (March 1977): 226-244.

Grofman, Bernard. "Some Notes on Voting Schemes and the Will of the Majority." *Public Choice* 7 (Fall 1969): 65-80.

Gudgin, Graham, and Peter J. Taylor. *Seats, Votes and the Spatial Organization of Elections.* London: Pion, 1978.

Haldeman, H. R., with Joseph DiMona. *The Ends of Power.* New York: Times Books, 1978.

Hansard Society Commission. *The Report of the Hansard Society Commission on Electoral Reform.* London: Hansard Society for Parliamentary Government, 1976.

Hardin, Russell. *Collective Action.* Baltimore: Johns Hopkins University Press, 1982.

Henderson, John M., and Richard E. Quandt. *Microeconomic Theory: A Mathematical Approach.* 2d ed. New York: McGraw-Hill, 1971.

Hillinger, Claude. "Voting on Issues and on Platforms." *Behavioral Science* 16 (November 1971): 564-566.

Hobbes, Thomas. *Leviathan.* 1651; New York: Liberal Arts Press, 1958.

Holsti, Ole R., Richard A. Brody, and Robert C. North. "Measuring Affect and Action in International Reaction Models: Empirical Materials from the 1962 Cuban Crisis." *Journal of Peace Research* 1 (1964): 170-190.

Howard, Nigel. *Paradoxes of Rationality: Theory of Metagames and Political Behavior.* Cambridge, Mass.: MIT Press, 1971.

Intriligator, Michael D., and Dagobert L. Brito. "Can Arms Races Lead to the Outbreak of War?" *Journal of Conflict Resolution* 28 (March 1984): 63-84.

——. "Formal Models of Arms Races." *Journal of Peace Science* 2 (Spring 1976): 77-96.

James, Marquis. *Andrew Jackson: Portrait of a President.* New York: Grosset and Dunlap, 1937.

Jaworski, Leon. *The Right and the Power: The Prosecution of Watergate.* New York: Reader's Digest Press, 1976.

Kadane, Joseph B. "On Divison of the Question." *Public Choice* 13 (Fall 1972): 47-54.

Kahan, James P., and Amnon Rapoport. *Theories of Coalition Formation.* Hillsdale, N.J.: Lawrence Erlbaum Associates, 1984.

Kearns, Doris. *Lyndon Johnson and the American Dream.* New York: New American Library, 1977.

Kelly, Jerry S. *Arrow Impossibility Theorems.* New York: Academic, 1978.

Kennedy, Robert F. *Thirteen Days: A Memoir of the Cuban Missile Crisis.* New York: W. W. Norton, 1969.

Key, V. O., Jr., with the assistance of Milton C. Cummings, Jr. *The Responsible Electorate: Rationality in Presidential Voting, 1936-1960.* Cambridge, Mass.: Harvard University Press, 1966.

Kilgour, D. Marc. "Anticipation and Stability in Two-Person Non-Cooperative Games." In *Dynamic Models of International Conflict,* edited by Michael D. Ward and Urs Luterbacher. Boulder, Colo.: Lynn Rienner, 1985.

——. "Equilibria for Far-sighted Players." *Theory and Decision* 16 (March 1984): 135-157.

Kuron, J. Interview. *Telos* 47 (1981): 93-97.

Lakeman, Enid. *How Democracies Vote: A Study of Electoral Systems.* 4th rev. ed. London: Faber and Faber, 1974.

Leontief, Wassily W., and Faye Duchin. *Military Spending: Facts and Figures, Worldwide Implications, and Future Outlook.* New York: Oxford University Press, 1983.

Levin, Murray. *The Alienated Voter: Politics in Boston.* New York: Holt, Rinehart and Winston, 1960.

Lijphart, Arend. *Democracies: Patterns of Majoritarian and Consensus Government in Twenty-One Countries.* New Haven, Conn.: Yale University Press, 1984.

Lorenz, Konrad. *On Aggression.* Translated by Marjorie Kerr Wilson. New York: Harcourt, Brace, and World, 1966.

Low-Beer, John R. "The Constitutional Imperative of Proportional Representation." *Yale Law Journal* 94 (November 1984): 163-188.

Lukas, J. Anthony. *Nightmare: The Underside of the Nixon Years.* New York: Viking, 1976.

Luterbacher, Urs. "Last Words about War?" *Journal of Conflict Resolution* 28 (March 1984): 165-181.

MacKay, Alfred F. *Arrow's Theorem: The Paradox of Social Choice.* New Haven, Conn.: Yale University Press, 1980.

Maddox, William S., and Stuart A. Little. *Beyond Liberal and Conservative: Reassessing the Political Spectrum.* Washington, D.C.: Cato Institute, 1984.
Majeski, Stephen J. "Arms Races as Iterated Prisoners' Dilemma Games." *Mathematical Social Sciences* 7 (June 1984): 253-266.
Mayston, David J. *The Idea of Social Choice.* New York: St. Martin's, 1974.
Meredith, James Creed. *Proportional Representation in Ireland.* Dublin, 1913.
Merrill, Samuel, III. "A Comparison of Efficiency of Multicandidate Electoral Systems." *American Journal of Political Science* 28 (February 1984): 23-48.
——. "Strategic Decisions under One-Stage Multi-Candidate Voting Systems." *Public Choice* 36 (1981): 115-134.
Meyer, Stephen M. "Verification and Risk in Arms Control." *International Security* 8 (Spring 1984): 111-126.
Miller, Nicholas R. "Logrolling and the Arrow Paradox: A Note." *Public Choice* 21 (Spring 1975): 107-110.
——. "Pluralism and Social Choice." *American Political Science Review* 77 (September 1983): 734-747.
Montagu, M. F. Ashley, ed. *Man and Aggression.* New York: Oxford University Press, 1968.
Moulin, Hervé. *The Strategy of Social Choice.* Amsterdam: North Holland, 1983.
Mueller, Dennis C. *Public Choice.* Cambridge: Cambridge University Press, 1979.
Murray, Alan. "Reapportionment: The Politics of N(N-1)." *Congressional Quarterly Weekly Report,* February 28, 1981, p. 393.
Muzzio, Douglas. *Watergate Games.* New York: New York University Press, 1982.
Myrdal, Alva. *The Game of Disarmament: How the United States and Russia Run the Arms Race.* New York. Random House, 1976.
——. "The International Control of Disarmament." *Scientific American,* October 1974, pp. 21-33.
Nagel, Jack H. *The Descriptive Analysis of Power.* New Haven, Conn.: Yale University Press, 1975.
Nash, John. "Non-cooperative Games." *Annals of Mathematics* 54 (1951): 286-295.
Newhouse, John. *Cold Dawn: The Story of SALT.* New York: Holt, Rinehart and Winston, 1973.
New York Times Staff. *The End of a Presidency.* New York: Bantam, 1974.
Niemi, Richard G. "Communications" (comment on S. J. Brams and P. C. Fishburn). *American Political Science Review* 79 (September 1985).
——. "The Occurrence of the Paradox of Voting in University Elections." *Public Choice* 8 (Spring 1970): 91-100.
——. "The Problem of Strategic Voting under Approval Voting." *American Political Science Review* 78 (December 1984): 952-958.
Niemi, Richard G., and Herbert F. Weisberg, eds. *Controversies in Voting Behavior.* 2d ed. Washington, D.C.: CQ Press, 1984.
Nincic, Miroslav. *The Arms Race: The Political Economy of Military Growth.* New York: Praeger, 1982.
Norton, Thomas J. "The Supreme Court's Five to Four Decisions." *American Bar Association Journal* 9 (July 1923): 417-420.
Nurmi, Hannu. "On the Strategic Properties of Some Modern Methods of Group Decision Making." *Behavioral Science* 29 (October 1984): 248-257.
——. "On Taking Preferences Seriously." In *Essays on Democratic Theory,* edited by Dag Anckar and Erkki Berndtson. Tampere, Finland: Finnish

Political Science Association, 1984, pp. 81-104.

____ . "Voting Procedures: A Summary Analysis." *British Journal of Political Science* 13 (April 1983): 181-208.

Pachter, Henry M. *Collison Course: The Cuban Missile Crisis and Coexistence.* New York: Praeger, 1963.

Palfrey, Thomas R. "Spatial Equilibrium with Entry," *Review of Economic Studies* 51 (January 1984): 139-156.

Pattanaik, Prasanta K. *Voting and Collective Choice: Some Aspects of the Theory of Collective Decision Making.* New York: Cambridge University Press, 1971.

Pillen, Herbert. *Majority Rule in the Supreme Court.* Washington, D.C.: Georgetown University, 1924.

Plott, Charles R. "Ethics, Social Choice Theory and the Theory of Economic Policy." *Journal of Mathematical Sociology* 2 (July 1972): 181-208.

____ . "Recent Results in the Theory of Voting." In *Frontiers of Quantitative Economics,* edited by M. D. Intriligator. Amsterdam: North Holland, 1971, pp. 109-129.

"Political Alienation in America." *Society* 13 (July/August 1976): 18-57.

Pursell, Carroll W., Jr., ed. *The Military-Industrial Complex.* New York: Harper and Row, 1972.

Rapoport, Anatol, and Albert M. Chammah. *Prisoners' Dilemma: A Study in Conflict and Cooperation.* Ann Arbor: University of Michigan Press, 1965.

Report of the Royal Commission Appointed to Enquire into Electoral Systems. London: HMSO, 1910, Cd. 5163.

Richardson, Lewis F. *Arms and Insecurity: A Mathematical Study of the Causes and Origins of War.* Pittsburgh: Boxwood, 1960.

Richelson, Jeffrey T. "The Keyhole Satellite Program." *Journal of Strategic Studies* 7 (June 1984): 121-153.

Riker, William H. "Arrow's Theorem and Some Examples of the Paradox of Voting." In *Mathematical Applications in Political Science,* edited by John M. Claunch. Dallas: Arnold Foundation of Southern Methodist University, 1965, pp. 41-60.

____ . *Liberalism against Populism: A Confrontation between the Theory of Democracy and the Theory of Social Choice.* San Francisco: Freeman, 1982.

____ . *Political Manipulation: The Art of Heresthetics.* New Haven: Yale University Press, forthcoming.

____ . "The Paradox of Voting and Congressional Rules for Voting on Amendments." *American Political Science Review* 52 (June 1958): 349-366.

____ . "Political Theory and the Art of Heresthetics." In *Political Science: The State of the Discipline,* edited by Ada W. Finifter. Washington, D.C.: American Political Science Association, 1983, pp. 47-67.

____ . *The Theory of Political Coalitions.* New Haven, Conn.: Yale University Press, 1962.

Riker, William H., and Peter C. Ordeshook. *An Introduction to Positive Political Theory.* Englewood Cliffs, N.J.: Prentice-Hall, 1973.

Romer, Thomas, and Howard Rosenthal. "Voting Models and Empirical Evidence." *American Scientist* 72 (September/October 1984): 465-473.

Rosen, Steven, ed. *Testing Theories of the Military-Industrial Complex.* Lexington, Mass.: Heath, 1973.

Rousseau, Jean-Jacques. *The Social Contract.* 1762; New York: Hafner, 1947.

Sarkesian, Sam C., ed. *The Military-Industrial Complex: A Reassessment.* Beverly Hills, Calif.: Sage, 1972.

Satterthwaite, Mark Allen. "Strategy Proofness and Arrow's Conditions: Existence and Correspondence Theorems for Voting Procedures and Social Welfare Functions." *Journal of Economic Theory* 10 (April 1975): 197-218.

Scammon, Richard M., and Ben J. Wattenberg. *The Real Majority: An Extraordinary Examination of the American Electorate.* New York: Coward, McCann and Geohegan, 1970.

Schattschneider, E. E. *The Semisovereign People: A Realist's View of Democracy in America.* New York: Holt, Rinehart and Winston, 1960.

Schelling, Thomas C. *Arms and Influence.* New Haven, Conn.: Yale University Press, 1966.

Scigliano, Robert. *The Supreme Court and the Presidency.* New York: Free Press, 1971.

Sen, Amartya K. *Collective Choice and Social Welfare.* San Francisco: Holden-Day, 1970.

———. "A Possibility Theorem on Majority Decisions." *Econometrica* 34 (April 1966): 491-499.

Shapley, L. S., and Martin Shubik. "A Method of Evaluating the Distribution of Power in a Committee System." *American Political Science Review* 48 (September 1954): 787-792.

Smith, John H. "Aggregation of Preferences with Variable Electorate." *Econometrica* 41 (November 1973): 1027-1041.

Snyder, Glenn H., and Paul Diesing. *Conflict among Nations: Bargaining, Decision Making, and Systems Structure in International Crisis.* Princeton, N.J.: Princeton University Press, 1977.

Sorensen, Theodore C. *Kennedy.* New York: Harper and Row, 1965.

Spanier, John W., and Joseph L. Nogee. *The Politics of Disarmament: A Study in Soviet-American Gamesmanship.* New York: Praeger, 1962.

Stavely, E. S. *Greek and Roman Voting and Elections.* Ithaca, N.Y.: Cornell University Press, 1972.

Stephenson, D. Grier, Jr. " 'The Mild Magistracy of the Law': U.S. v. Richard Nixon." *Intellect* 103 (February 1975): 288-292.

Still, Edward. "Alternatives to Single Member Districts." In *Minority Vote Dilution,* edited by Chandler Davidson. Washington, D.C.: Howard University Press, 1984, pp. 249-267.

Straffin, P. D., Jr., M. D. Davis, and S. J. Brams. "Power and Satisfaction in an Ideologically Divided Voting Body." In *Power, Voting, and Voting Power,* edited by Manfred J. Holler. Würzburg, West Germany: Physica-Verlag, 1982, pp. 239-255.

Straffin, Philip D., Jr., "The Bandwagon Curve." *American Journal of Political Science* 21 (November 1977): 695-709.

———. *Topics in the Theory of Voting.* Boston: Birkhäuser, 1980.

Straffin, Philip D., Jr., and Bernard Grofman. "Parliamentary Coalitions: A Tour of Models." *Mathematics Magazine* 57 (November 1984): 259-274.

Sykes, Lynn R., and Jack Evernden. "The Verification of a Comprehensive Nuclear Test Ban." *Scientific American,* October 1982, pp. 47-55.

Szafar, Tadeusz. "Brinkmanship in Poland." *Problems of Communism* 30 (May/June 1981): 75-81.

Taylor, Michael. *Anarchy and Cooperation.* London: Wiley, 1976.

Taylor, Michael J. "Graph-Theoretical Approaches to the Theory of Social

Choice." *Public Choice* 4 (Spring 1968): 35-48.

Taylor, Peter J., and Ronald J. Johnston. *Geography of Elections.* London: Penguin, 1979.

Tierney, John. "The Invisible Force." *Science '83,* November 1983, pp. 65-78.

Totenberg, Nina. "Behind the Marble, Beneath the Robes." *New York Times Magazine,* March 16, 1975, pp. 15ff.

Udis, Bernard, ed. *The Economic Consequences of Reduced Military Spending.* Lexington, Mass.: Heath, 1973.

von Neumann, John, and Oskar Morgenstern. *Theory of Games and Economic Behavior.* 3d ed. Princeton, N.J.: Princeton University Press, 1953.

Wagner, R. Harrison. "The Theory of Games and the Problem of International Cooperation." *American Political Science Review* 78 (June 1983): 330-346.

Warren, Charles. *The Supreme Court in United States History,* vol. 1. Boston: Little, Brown, 1924.

Watt, Richard M. "Polish Possibilities." *New York Times Book Review,* April 25, 1982, pp. 11, 19.

Weisberg, Herbert F., and Richard G. Niemi. "Probability Calculations for Cyclical Majorities in Congressional Voting." In *Probability Models of Collective Decision Making,* edited by Richard G. Niemi and Herbert F. Weisberg. Columbus, Ohio: Charles E. Merrill, 1972, pp. 104-131.

Wittman, Donald. "Candidate Motivation: A Synthesis of Alternative Theories." *American Political Science Review* 77 (March 1983): 142-157.

Wohlstetter, Albert. "The Delicate Balance of Terror." *Foreign Affairs* 37 (January 1959): 209-234.

Woodward, Bob, and Carl Bernstein. *The Final Days.* New York: Simon and Schuster, 1976.

Wright, James D. *The Dissent of the Governed: Alienation and Democracy in America.* New York: Academic, 1976.

The Writings: Kethubin. Philadelphia: Jewish Publication Society of America, 1982.

York, Herbert F., and G. Allen Greb. "Strategic Reconnaissance." *Bulletin of Atomic Scientists,* April 1977, pp. 33-42.

Zagare, Frank C. "A Game-Theoretic Analysis of the Vietnam Negotiations: Preferences and Strategies 1968-1973." *Journal of Conflict Resolution* 21 (December 1977): 663-684.

——. "A Game-Theoretic Evaluation of the Cease-Fire Alert Decision of 1973." *Journal of Peace Research* 20, no. 1 (1983): 73-86.

——. *Game Theory: Concepts and Applications.* Beverly Hills, Calif.: Sage, 1984.

——. "The Geneva Conference of 1954: A Case Study of Tacit Deception." *International Studies Quarterly* 23 (September 1979): 390-411.

——. "Limited Move Equilibria in 2 x 2 Games." *Theory and Decision* 16 (January 1984): 1-19.

——. "Nonmyopic Equilibria and the Middle East Crisis of 1967." *Conflict Management and Peace Science* 5 (Spring 1981): 139-162.

——. "Toward a Reformulation of the Theory of Mutual Deterrence." *International Studies Quarterly* 29 (June 1985).

Zinnes, Dina A., and John V. Gillespie, eds. *Mathematical Models in International Relations.* New York: Praeger, 1976.

Index